C0 CGF 615

RELIGION AND I
NORTHERI

Religion and Employment in Northern Ireland

DAVID EVERSLEY

SAGE Publications
London · Newbury Park · New Delhi

First published 1989

SAGE Publications Ltd
28 Banner Street
London EC1Y 8QE

SAGE Publications Inc
2111 West Hillcrest Drive
Newbury Park, California 91320

SAGE Publications India Pvt Ltd
32, M-Block Market
Greater Kailash – I
New Delhi 110 048

British Library Cataloguing in Publication data

Eversley, David, *1921–*
Religion and employment in Northern Ireland.
1. Northern Ireland. Catholics. Employment.
Equality of opportunity
I. Title
331.13′3

ISBN 0–8039–8203–8

Library of Congress catalog card number 88–63340

Typeset by Fakenham Photosetting Ltd, Fakenham, Norfolk
Printed in Great Britain by Billing and Sons Ltd,
Worcester

Contents

Preface and Acknowledgements

This report was commissioned by the Fair Employment Agency for Northern Ireland in 1981. Though much of the information, especially about total population size, the proportion of Roman Catholics, employment and industrial structure and educational attainments, all refer to 1981, the Census year, many of the series based on current monitoring of the Northern Ireland labour market have been brought up to date and the overall unemployment position to which the report relates is that of April 1985. Unemployment in 1981 stood at just under 17 percent, and by 1985 had risen to 21 percent. This rise was somewhat less than that experienced in the United Kingdom as a whole, and a good deal less than in South East England. The main rise in Northern Ireland unemployment took place between 1979 and 1981, and levelled out thereafter. One therefore has some grounds for the belief that the main factors analysed in this report were in evidence by the time of the 1981 Census, and those sections which analyse developments since that time do not introduce any fundamentally new factors.

Nevertheless, it is regretted that the publication of the report has been delayed. In part this is due to the fact that certain other research projects (for example in the field of education) were still awaiting completion, and it was necessary to await these findings to complete the analysis of causes of differential employment market developments.

The generous cooperation of the agencies and individuals mentioned below has meant that we were inundated with material. In some cases we had, however, to wait until the end of 1984 before important statistics became available (for example unpublished Census information), and there seemed no point in completing the text and drawing up tables until all the relevant material was delivered.

The text is the sole responsibility of the author. If, throughout the text, the authorship appears in the plural, this indicates that all the work has in fact been collaborative, and that the findings have at all stages been discussed with the experts in Northern Ireland. Most of the issues arising from the Census undercount and non-response rates, in particular, were resolved jointly with Valerie Herr of Berkeley, who was the coauthor of a preliminary study on the Roman Catholic population in 1981 (Eversley and Herr, 1985).

My first debt is to the Fair Employment Agency for Northern Ireland, especially its Chairman, R. G. Cooper, and to its staff members, notably P. Sefton and Eileen Lavery. The Agency was instrumental in making initial contacts for me, in procuring unpublished statistics from many government departments and, in the early stages, undertaking some of the laborious computing work involved in translating raw data into percentages and tabulating the results in an easily assimilable form. Eileen Lavery in particular made herself available throughout the currency of the project to discuss difficulties and to make suggestions as to the method to be employed, as well as the significance of the outcome. She also supervised internal staff, provided the base maps for the text figures, and supervised ancillary staff in Belfast.

Robert Osborne of the University of Ulster, apart from providing me with advance texts of his publication in the field of education in relation to the labour market, was also very helpful in discussing some of the methodological issues, as well as the findings. Our debt to his work will be apparent throughout.

The Northern Ireland Census Office, mainly in the person of Charles Daly, was unfailingly obliging in the matter of unpublished Census information. The whole of the sections based on the Census table groups 8, 9 and 10 (employment, industry and education) are based on the Census Office's special tabulations and excerpts.

In England, my chief debt must be to Valerie Herr, who worked with me on the religion report (Eversley and Herr, 1985) but also discussed many other issues of statistical methodology with me while she was spending a year in Cambridge. Her long experience of extracting social demographic data from imperfect sources was of great benefit.

For most of the project's lifetime, Jan Trow-Smith was in charge of the office, storing data, producing text and tables, and conducting negotiations with source offices. My main debt for the complex calculations by which raw data were transformed into meaningful text and tables is to Nadine Wylie and Juliette Thorp. The thankless task of editing the manuscript, the tables and the figures was entrusted to Elizabeth Parker and Barbara Pennell. Andrew Hoddinott and Nadine Wylie drew most of the figures. The manuscript was typed by Jan Trow-Smith. Office Overload (Helen Baron and Caroline Wilkinson) produced the more complex tables from handwritten data. My thanks go too to Linda Haley who helped with proof reading and indexing.

In the list of Northern Ireland government agencies which follows, it would be invidious to single out, in each case, the link person who was instrumental in providing data and answering technical queries. No effort was spared to mobilize often very inaccessible data for the purposes of this exercise.

The Policy Planning Research Unit within the Department of Finance and Personnel assisted us by discussions on methodology and by making available to us advance printouts based on the results of the 1983 round of the Continuous Household Survey. Keith Wilson-Davies and later Kevin Sweeney were instrumental in interpreting the data for us. In this connection it is proper to acknowledge the positive response to our appeals for assistance to two successive ministers in the Northern Ireland Office, the Earl of Gowrie and Rhodes Boyson, when the future of the Continuous Household Survey, and later the analysis of the survey data, seemed to be in jeopardy. Senior civil servants in the Department of Finance and Personnel also took a benevolent interest in what we were doing and opened the way forward for us on a number of occasions.

The departments involved in the provision of data were as follows:

Department of Economic Development (DED)
Department of Health and Social Services (DHSS)
General Register Office (GRO)
Census Office
Department of Education for Northern Ireland (DENI)
Department of the Environment (DoE)
Department of Finance and Personnel, Policy Planning Research Unit (PPRU)
Northern Ireland Housing Executive (NIHE)

The Fair Employment Agency for Northern Ireland not only defrayed the entire cost of this research, as already detailed, but also assisted publication in book form by a grant-in-aid.

David Eversley
Cottered

Note on Statistics

To avoid overloading this report with detailed statistics, especially those relating to local (district-level) experience in the employment market, the full tables which furnish the evidence for the conclusions in the text are available separately.

These Additional Tables were prepared by the same team which compiled the text tables, in England, and the word processing was carried out in Belfast by Betty Haseley and her team. Thanks are due to them also.

The volume of Additional Tables may be obtained upon application to the Fair Employment Agency for Northern Ireland, Andras House, 60 Great Victoria Street, Belfast, BT2 7BB.

Throughout the text, references to the Additional Tables in this supplementary volume are prefixed by the letter A. A full list of these tables is given in Appendix D.

List of Tables

List of Figures

Introduction

The Fair Employment Agency for Northern Ireland, which commissioned this research, is a government-financed but independent body working along lines closely resembling the activities of the Equal Opportunities Commission and the Commission for Racial Equality. That is to say, it attempts to ascertain the extent of discrimination, in the labour market, against any group in the community, but in particular the Roman Catholic minority. It investigates complaints, gives advice and publishes its own findings.

In Northern Ireland we can identify two polarized views about the differences in unemployment rates as between the two main religious groups; first, that this is wholly, or at any rate in large part, due to discrimination by employers; and secondly, that the higher Roman Catholic unemployment rates are largely caused by that group itself, mainly because of its higher fertility, but also because of areas where its members choose to live, or because of the socio-economic structure of the districts of high unemployment. This second view has been authoritatively summed up by Paul Compton of the Queen's University of Belfast:

> The analysis presented in this paper has been of a macro-nature. It has produced evidence that, while some unfairness in job allocation may exist, it is structural imbalances generated by factors specific to the Roman Catholic community, such as higher rates of population growth, lower social status, larger families, and a divergence between geographical distribution and the location of jobs, that account for a considerable part of the disparity between the respective denominational rates of unemployment, and hence contribute to inequality of job opportunity. Such factors are in intimate connection with each other as well as with unemployment through a complex system of linkages set in an historical and political context. For instance, social class and denomination are both determinants of family size and the birth rate, although in Northern Ireland denomination is the more important of the two; the rate of natural increase influences the level of emigration; the informal networks of job recruitment, considered of such significance by the Fair Employment Agency, can only work to the advantage of Protestants through demographic mechanisms; and so on. It is the contention here that equality of job opportunity for Roman Catholic *vis-à-vis* Protestant is a meaningless concept while these fundamental parameters of a demographic and geographical nature remain so unequal. Moreover, so long as existing circum-

> stances prevail, equality for Roman Catholics can only be achieved by long-term Protestant inequality because the structural imbalances generated by high population growth, large families, etc. are perpetuated from generation to generation. In other words, the only effective way to guarantee improvement in the relative position of Roman Catholics in Northern Ireland is through the encouragement of fundamental change in certain of the innate features of that community. Acceptance of the desirability to bring Roman Catholic family size and rate of growth closer to the national and European average would be an important step along the road to greater equality. (Compton, 1981)

To quote this passage is not to imply that this research report has been organized as a response to the views voiced by Compton. The problem is much more complex than this brief quotation would seem to imply. Nevertheless, it forms a starting point to the investigation in the sense that it mentions some of the main areas we need to investigate.

Method

The method used here is essentially designed to *explain* as much of the differential in local (or community) unemployment rates as we can, by reference to demographic factors in the first place, but also by looking at the structure of local labour markets. We shall also investigate the educational and training opportunities open to the local labour force entrants.

In many investigations undertaken in the last few years, the question of discrimination in the housing market has also been studied. It has been assumed that the inability of minority workers to obtain jobs may have been due to discrimination in housing allocation. This line of inquiry did not prove fruitful in the context of our labour market analysis and it will not be separately presented here. However, there remains the context of the more general problem of residential segregation, and access to workplaces. This means that housing is, in Northern Ireland, part of a much larger socio-political difficulty in which the operation of housing allocation plays a part but does not have all that much explanatory value.

Accordingly, we shall concentrate mainly on the demographic structure of the different areas of Northern Ireland, in terms of age, sex and educational attainment, and on other characteristics of the population which may throw light on their economic activity rates, unemployment experience and mobility.

Unemployment was never equally distributed between all areas, in Northern Ireland any more than in the rest of the UK. For the whole of Ireland, the excess of potential working population over the capacity of the country to provide an income has been an endemic problem since the middle of the nineteenth century. The history of the Great Famine and its aftermath, and the connection of the carrying capacity of the land with landholding patterns and attempts to reform these, has been amply chronicled. Emigration was the short-term remedy, and remained the main safety valve until quite recent times (Jackson, 1963; Compton, 1974; Cullen, 1972: Chapters 5 and 6). Migration similarly served to restore a measure of equilibrium to other labour markets in the UK where either industrial change or agricultural innovation created surplus population. How far overseas migration accounted for local population reduction, and how far this was accomplished by internal movement, especially into the new areas of industrial growth, need not concern us here.

Population shifts at a regional level were usually sufficient to satisfy the demand for labour, especially in new industries. For the UK as a whole, however, long-term regional differentials in employment and incomes began to appear in the 1930s, and were only briefly reduced during the war years. From the early 1970s onwards, regional unemployment differentials again began to grow, and Northern Ireland was, at all times, the worst-hit region of the UK. This is not the place to discuss the merits or otherwise of the regional policies pursued by successive UK governments, or the implications of virtual discontinuation of these attempts in recent years. Although many analysts regard Northern Ireland as being nothing more than an extreme case of the general malaise of the industrial world (and the British example in particular) we shall, in this report, not discuss this view, but concentrate on the local aspect of the question which hinges substantially on the matter of religious affiliation.

At the end of the chapters dealing with differentials in local employment and industrial structures in Northern Ireland, with education, and to a lesser extent with housing and car ownership, we shall face the fact that these variables do not in fact wholly explain differences in unemployment rates (and, by implication, in household incomes). At this point we shall then have to face the major social and political question: if people do not obtain employment where they live, how much of this may be due to discrimination against them? If they do not move, how far can this be said to be due to actual experience of discrimination or threats to personal feelings of security, or due to the fear that discrimination or harassment makes a move to more favourable labour markets hazardous, or even impossible?

We shall be left with the conclusion that a good deal of the differential unemployment in certain areas is indeed due to the inherent local labour market structure, which many years of governmental policies have done little to change. We shall also assume that whatever mechanisms, in the past, kept this differential within bounds, there are new factors which have now increased its significance. Under the heading of 'inherent' we shall naturally investigate the demographic structure as well as the employment market.

Some Limitations of the Inquiry

The inquiry into the social demography of the Northern Irish labour market began in 1981. At that time no information from the 1981 Census was available, and the basis of employment accounting was undergoing a process of change. Industrial and occupational classifications were also being changed. We knew therefore that long series of comparable data would be hard to obtain.

When the 1981 Census results did begin to come in, it was immediately apparent that the Census suffered from grave deficiencies, and would have to be used with great caution. As in the rest of the UK, changes in local government boundaries made comparisons with 1971 extremely difficult. In addition there was both underenumeration and a high level of non-response to the voluntary question on religion. This particularly strong resistance encountered to the question of religious affiliation makes tables which purport to compare religious structure in 1971 and 1981 highly suspect. By 1983 the Continuous Household Survey (CHS) had been launched by the Policy Planning Research Unit (PPRU) and the first results became available in 1984, but in insufficient quantity, detail and time series to add very much to our previous findings.

Areal Definitions

The country which is officially called Northern Ireland is also referred to, in the literature generally and also in this report on occasions, as 'Ulster', or simply 'the Province'; neither of these terms has any legal significance, and 'the Six Counties' is now an obsolete term.

The difficulty begins when we try to subdivide the country. The *counties* lost their status as such during the 1973 reorganization, though they still figure in postal addresses and in much informal usage. The *districts* subsequently created vary widely in area and population, from under 30,000 to nearly 400,000 (Figure 1), and though they will be used in some of the analytical tables (mainly in the volume of Additional Tables) they are far too small for most purposes. When we are dealing with the division of the population, for instance into industrial and occupational groups, the resulting numbers at district level are often so small that it would be misleading to base any conclusions on them.

Officially (that is, as far as the Census Office is concerned, and in some other government publications) the country has four *regions* (see, for example, Northern Ireland Census 1981, *Preliminary Report*: v) but because the Belfast metropolitan area, for instance, is divided between the Eastern and Northern regions, it will be seen that this division is not very useful.

For employment purposes, the country is divided into *travel-to-work areas* (TTWAs) which are a little more meaningful. The definition of these areas was changed after the 1981 Census, as the result of the pattern of work journeys revealed by analysis of that year, and the new areas were brought into use in 1984. Comparisons of statistics before and after that date are therefore not possible. However, during the crucial period when unemployment was rising rapidly, the TTWAs remained unchanged (for definitions see Figure 2). Until 1984, TTWAs were divided into employment service office areas (ESOs) and these were useful for some purposes, though their location was governed by administrative convenience rather than economic reasons. They do not help us much with our labour market analysis.

From 1985, the Belfast travel-to-work area was greatly expanded by the Department of Economic Development (DED) to take account of the realities revealed in the Census workplace and transport to work volume. The new Belfast TTWA now includes Antrim and Down. Our own area group I (Belfast) does not follow this pattern: we have grouped Antrim with the Northern districts and Down with the Southern. Because TTWAs were unchanged for most of the years covered by this report, we have not adopted the new boundaries. For many other purposes we thought it appropriate to keep to our four area groupings. One reason is that the DED's new TTWA Belfast is such a large entity that it swamps even more the employment differentials which exist between areas. (On the new boundaries it comprises 58.5 percent of the Northern Ireland labour market, with no other TTWA having as much as 10 percent of the whole market and several having under 2 percent.)

In *education* the country has five Education and Library Boards which are much used by the Department of Education for Northern Ireland (DENI) for its own returns (see Figure 3). Other types of boundary exist for *housing* (the Northern Ireland Housing Executive has its own four areas) and for *health* services. We could add electricity supply areas, water boards and many others irrelevant for our own analysis. Unfortunately, many of these boundaries do not

Figure 1 *Northern Ireland local government districts*

coincide with district council boundaries, or former counties (compare Figures 1–3).

There is no single and simple way to subdivide the Province which would serve every type of analysis. This limits the validity of some of the comparisons made. We must remember, however, that we are dealing with an area where journeys to work, to school or further education, or to hospital, are never as long for the great majority of people as they are in thinly settled rural or upland areas elsewhere in the British Isles. Thus distance between home and workplace, or service facility, is a less crucial variable than other causes of dissimilar provision.

Some of these problems are familiar in the rest of the UK, but are simplified there by the continued use of standard regions. These first appeared in the 1960s and, for most purposes of economic analysis, population change and other pertinent matters, have proved useful for interregional and time trend analyses. The UK government statistics (now called *Regional Trends*) represent an element of continuity. Northern Ireland is, of course, smaller than the smallest British region (East Anglia), and therefore further subdivision becomes more difficult.

In popular parlance, Northern Ireland has been supposed to be

Figure 2 *Northern Ireland travel-to-work areas to 1984 (and employment service office areas)*

divided into two main areas: 'east of the Bann' and 'west of the Bann'. Convenient as this sounds, it has no real validity in either social or economic analysis, and we do not use it here. There is rather more point in dividing the Province into a *core* area and the *periphery* – the first meaning, of course, Belfast and its immediate hinterland, and the rest containing all other districts. This division allows us to subsume under the 'peripheral' heading areas like North Antrim, South Armagh and South Down, which are in the east of the country but have all the characteristics of an economically peripheral region. Unfortunately, however, this useful concept cannot be easily translated into the areas we have to use for our labour market analysis. It is often claimed that the 'core' is mainly Protestant, and the 'periphery' Catholic. There is some truth in this, but again the generalization involved cannot serve except as a convenient summary of a much more complex situation.

We have therefore in this report divided the Province into four most-purpose *area groups* (of pre-1984 travel-to-work areas). We

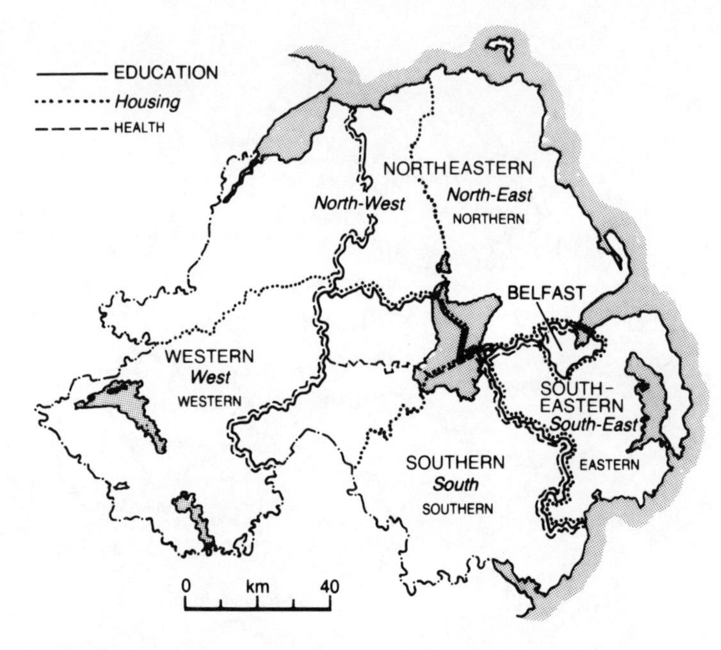

Figure 3 *Northern Ireland education, health and housing boundaries*

believe these correspond sufficiently to a number of important socio-economic subdivisions to serve as a basis for a multifaceted analysis. Our maps are also based on these four area groups (see Figure 4), and they are so referred to in our table headings. (For a list of districts in each area group, see Appendix A.) There is nothing normative about these groupings, and we are well aware that there is a good deal of movement (for example, of pupils in secondary schools, and even more so of students in further education) across the demarcation lines, and even across the national boundary into and out of the Republic. (Not to speak of the diocesan and parish boundaries which, as we showed in our religion report, are unrelated to the 1922 border: Eversley and Herr, 1985.) Certainly people travel to work both ways, not only between districts and area groups, but again over the national border.

The Statistical Sources

All Northern Ireland statistics need to be treated with even more caution than one would exercise when using the statistical returns of

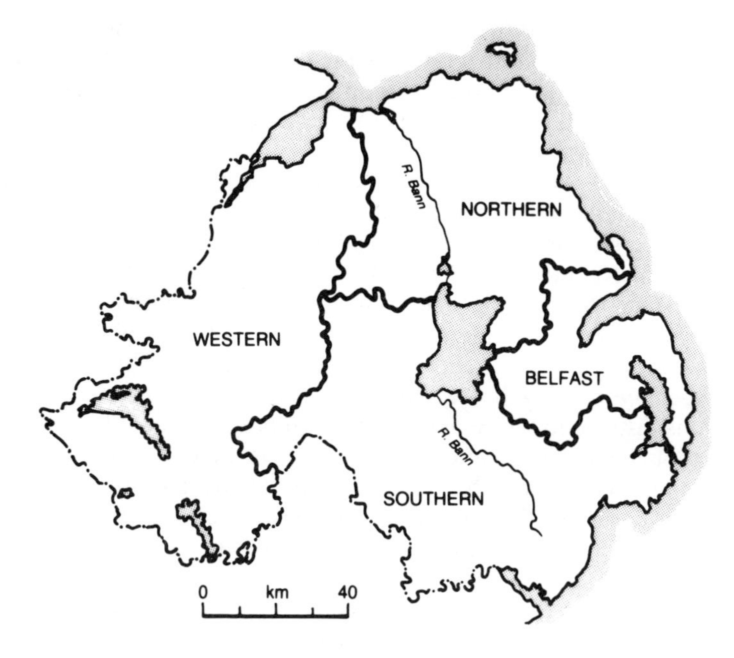

Figure 4 *Northern Ireland base map, four area groups and line of River Bann*

any country. This is partly because of the voluntary question on religion in the Census, and partly because the relationships between government and groups in the community are often strained, and respondents filling out forms and questionnaires often feel that their answers may have unwelcome policy repercussions. The notes which follow give only a general indication of the assumptions which have been made.

First, no attempt will be made to reopen the question of the exact size of the population in 1981, its local distribution or its age structure. At the time of writing, the best estimate available was that published in 1984 by the Registrar General for Northern Ireland (General Register Office, 1984) in which the total is given as about 1,562,000 persons. All the main analyses in this volume have been cross-checked with that figure. This publication is the only one on which sex, age and local area distributions could be based, other than the obviously defective Census. Since the completion of the main report there have been several revisions, the most important being a new 'best estimate' of underenumeration by the Policy Planning Research Unit (PPRU) of the Department of Finance and Personnel (PPRU, 1986). This publication arrives at a total population of about

1,532,000. This is a difference of about 2 percent, and though the overall difference conceals rather larger differences at district and particularly ward level, the revision does not affect the general analysis of this study. It will have to be taken into account when looking at intercensal migration.

Secondly, we shall use the estimates of the relative size of the Roman Catholic and 'other' populations as set out in our previous report (Eversley and Herr, 1985) and not enter otherwise into the debates which continue on this matter. For some purposes it will still be necessary to use the religion tables of the 1981 Census report (imperfect as these are), largely in order to determine certain relative orders of magnitude rather than absolute figures. This will be explained more fully below. Since that time, other estimates of religious distribution have appeared which differ in detail from our own, though they do not affect the overall percentage of Catholics in the population. These other research reports appeared too late to enable us to discuss their merits or otherwise, even if such a debate had been necessary for the purposes of the present study.

Thirdly, we shall in this report quote only certain rather rough estimates of relative fertility for the main religion groups, as we did in Eversley and Herr (1985). Apart from that, we follow the main findings now made available in the last published Fertility Surveys (Compton et al., 1985).

Fourthly, there are a number of other areas where we are heavily dependent on poor records, guesswork and indirect estimates, for example movement between the Irish Republic and the North, or even movement between Ulster and the rest of the UK. Statistics relating to self-employment, part-time work and unregistered work by apparently economically inactive persons are grey areas in Northern Ireland that are at least as serious as elsewhere, if not more so. These difficulties will be noted where required. Some information on part-time working is available from the Census of Employment, and the Labour Force Survey throws light on some of the other grey areas.

Apart from these limitations of the scope of this report, some general observations on the nature of the sources are requisite.

We begin with the census. It is generally acknowledged that the 1981 Census underrecorded the population more seriously than had been the case in 1971, that this underenumeration was very unequally distributed through the country, and that, in addition to the failure of about 20 percent of the respondents to state their religious affiliation (double the proportion in 1971), other parts of the Census schedules may have been inadequately answered. These matters are the subject of at least one official publication (Morris et al., 1985) and a number of research monographs from academic sources (Compton, 1981; Compton, 1985a). We have here accepted the Registrar General's final estimate for the 1981 midyear population, published in 1984 after taking account of all other possible sources of information, as our baseline (General Register Office, 1984). The Registrar General's figures, like the amended ones produced by the PPRU, are confined to district totals and a breakdown by age and sex. They cannot therefore be directly compared with figures derived from Census tabulations because we cannot assume that the shortfall for any particular category (in the fields of economic activity, housing, socio-economic groupings, or education) are proportional to the underenumeration of the whole district population. Each analysis in this study must therefore be treated, to some extent, as a separate exercise: the base figure to which it refers has to be the Census total, or an educational return, or a set of travel-to-work area statistics. The extent and possible bias of any shortfall will vary greatly according to the type of source. Thus it seems likely that more households in public rented housing, especially in the two major city regions, were omitted than owner-occupied suburban or rural property.

If, as we tried to demonstrate in our report on religious affiliation (Eversley and Herr, 1985), the non-respondents to that question should be allocated roughly in the proportion of 55:45 nationally, and if, as seems likely again, Roman Catholic workers are overrepresented in the lower socio-economic status groups, then the non-respondents in those groups in the economic activity tables cannot be allocated in the same proportion as all workers. The upshot of this is that precisely in those fields where we would most like to know the religious affiliation of individuals in order to come to a view about their educational attainments, about their occupational status, their tenure and so on, we should be prejudging the issue if we allocated the non-respondents (or the 'not stated' in the religion tables) in exactly the same proportion as either the respondents, or the overall ratio for the 'not stated'. Nevertheless, as we shall show, even the Census religion tables can sometimes be used for analysis provided other rules are observed.

Other Northern Irish data sources are not as suspect as the Census. The allocation of insured population to industries, the results of the annual Labour Force Survey, the analysis of the unemployed collected by the Department of Economic Development, the educational enrolment and attainment statistics, are all much more reliable – though here, as elsewhere, they always have to be used with caution. The preliminary results of the analysis of the Continuous Household Survey (CHS) do not yet allow a regional breakdown (though urban and non-urban populations can be differentiated). However, the extent to which one can rely on the results of the CHS is debatable. Its authors claim that refusal to participate, or to answer certain questions, cannot be linked either with political or with religious affiliation of the respondents, and that the published tables (PPRU *Monitors*, no. 1/84 onwards) are as reliable as similar Household Survey tables prepared in Great Britain. This matter cannot be argued here, and we have utilized the CHS rather selectively so as to shed light on questions, in the main, where any bias in non-response or incomplete responses is not likely to affect the outcome (for example, intention to move; see Chapter 2).

The non-congruence of various types of regional boundaries, already alluded to, often makes comparisons difficult, as does the problem of accounting periods: calendar years, school years, financial years, intercensal decades, and midyear estimates. When the old standard industrial classifications (SICs) were changed, after 1980, to a new system of industry divisions, there was a discontinuity which we have tried to overcome in the relevant tables of industrial distribution. However, these procedures are obviously subject to criticism from other investigators. Our summary of industries in the new classification (in five groupings) is necessarily arbitrary, as is our compression of school-leavers' educational attainments. In work of this kind there is always the problem of either showing data in excessive detail, or making assumptions about subdivisions which may not be shared by others. But in every case we either print the full data in the Additional Tables, or refer the user to the original tables.

In a few cases, we were unable to reconcile tabulations coming from one source with those coming from another (for example, as regards the number of occupied dwellings, a matter discussed in our religion monograph: Eversley and Herr, 1985).

In a number of cases, as we shall point out, even carefully collected figures from official sources will not serve to sustain a detailed and meaningful analysis. Thus, when unemployment is analysed by industry, those who are registered and claiming benefit may never have worked in any industry, and thus come into the ever-growing category of 'not elsewhere specified'. As higher proportions of youngsters fail to find jobs when leaving full-time education, industrial breakdowns of the unemployed assume mostly historical significance. Broad categories like 'married women workers in part-time employment' can mean anything from a science teacher working eighteen hours a week and earning between £100 and £150, to a part-time hospital cleaner earning £20. In other words, all tables, figures and graphs must be read in conjunction with the explanatory text; they are often meaningless by themselves.

In view of all these difficulties, including the unreliability of so much of the source material, we have tried to draw no inferences from the data at our disposal except where *differences* were large enough to be robust; that is, they would still be significant even if some of the base material were seriously flawed. Thus, if differences in the characteristics of the religious affiliation groups, as recorded in the Census, were large enough to be valid even if, let us say, the division of the 'not stated' in a particular case were not 55:45 (as we assume overall) but 25:75, then we can have some confidence in the reliability of the material.

These caveats are commonplace in the analysis of social and economic data from official statistics. There is, however, in this case a more important reason for our confidence in the material. All the indications are that the underenumeration, or non-response rate to certain questions, was biased overall so that relatively more Roman Catholics did not respond, relatively more public sector tenants, relatively more persons in the lower socio-economic groups, relatively more unemployed, and relatively more younger people with children.

Therefore, any statement we make about local or intergroup differences will err on the side of *understating* the disadvantages of the minority group. If we had been able to include the whole 'shadow' population (those who do not appear in the Census, in the Continuous Household Survey, on registers of insured or unemployed workers, and on electoral lists, and at best only on school attendance rolls) there would have been, we feel certain, predominantly Roman Catholics in the most disadvantaged categories. We found underenumeration to taper off with age, so that fewer over 50s went unrecorded, and relatively fewer in the higher socio-economic groups. The full-time employees in the Census were not too widely different from the known insured population, especially amongst males. On the other hand, non-response to the religion question in the Census was not so clearly stratified, with significant refusals to answer this question in middle-class groups in, for instance, the better residential suburbs of Belfast.

These considerations are necessary as a background to the analysis which follows.

Summary of the Argument and Procedure

Background to the Investigation

Northern Ireland has consistently had higher rates of unemployment than the rest of the UK. Going back no further than the 1960s, when full employment was taken for granted, the Province stood out with its much higher rates. In the rest of the UK the unemployment rate stood at a rather artificial figure of around 2 percent until it began to rise very slowly in the 1970s. In contrast, Northern Ireland recorded a figure of as much as 9.3 percent in 1958, and after that generally stood at about three times the British level. However, when the recession began to show itself, Northern Ireland was not at first seriously affected, and for a short time unemployment actually declined. As late as 1974 the rate was 6.1 percent and, although the percentage rose after this, it did so less rapidly than elsewhere in the UK, so that after the early 1970s the Northern Ireland figure was almost always less than twice as high as that for the rest of the UK.

In the mid 1980s the relativities still apply, though it is no longer possible to make exact comparisons. Changes in benefit rules, and the introduction of temporary employment schemes which especially take young people from the register, make it hard to say exactly how many people are unemployed: it may well be more than 15 percent of the population wanting to work in the UK as a whole, and over 25 percent in Northern Ireland. Equally, local differences are nothing new; significant areas of the rest of the UK now have 25 percent or more adult unemployment, and the figure in Northern Ireland may be locally as high as 50 percent. At this sort of level, exact percentages are not very meaningful: so many potential workers must be discouraged from registering at all, and the recorded figure merely relates to claimants who state that they are available for work. Many more people will have retired prematurely from the labour market, and others will be registered in the Census as chronically sick; these do not show up at all in current employment statistics.

The figure that has risen dramatically, in Northern Ireland as in the rest of the UK, is that of the adult long-term (over one year) unemployed. Equally, the proportion of school-leavers who fail to find a job has increased considerably more than unemployment in general, after allowing for the one-year temporary schemes which mean that these young people do not join the register until some time after completing their education.

What then are the special factors now affecting unemployment in Northern Ireland? First, the traditional safety valve of emigration has been all but closed. We shall in Chapter 2 discuss the possible direct effect of the UK recession (especially in the construction industries) on emigration from Northern Ireland. The Irish Republic, too, has experienced a deep recession following the temporary boom in the early 1970s. The annual flow of 2,000–2,500 persons from the North to the South slowed down considerably in recent years (PPRU, 1986: Table 6). Overseas migration is now a mere trickle, even allowing that the tiny sample statistics of the International Passenger Survey might be used as a good basis for any measurement. There are still considerable numbers of Irish-born workers in Great Britain (Office of Population Censuses and Surveys and Registrar General Scotland, 1983). Whether or not these people will stay in Great Britain cannot be forecast: until the most recent years there was still net migration from Northern Ireland to Great Britain, and the factors which contribute to the decision to stay or move, in either direction, are too complex to permit any projection.

In the UK as a whole, public sector employment peaked some time in the early 1970s, and then stabilized and eventually went down. In Northern Ireland the governmental economy measures have not so far been quite so drastic but, as we shall show, growth has also stopped in the Province in this sector. Future chances of full-time employment seem dim, especially as the public sector workforce is relatively young, thanks to the rapid expansion ten to fifteen years ago.

As regards manufacturing, Northern Ireland again experiences the same general downturn as the rest of the UK, but the impact has been more severe for certain locations where a single firm provided almost the entire demand for adult semiskilled or skilled labour – a situation comparable with the steel towns of Great Britain, and the centres of motor vehicle manufacturing. Again, the running down of regional aid has not been as drastic in Northern Ireland as in Great Britain, but the maintenance of assistance levels has scarcely affected the speed of collapse.

That much at any rate is agreed, and forms the substance of numerous monographs, many of them emanating from the Northern Ireland Economic Council (Northern Ireland Economic Council, 1985), the Department of Applied Economics in Cambridge, and other research centres. In 1985 a new major centre for economic policy research was endowed in Belfast, as well as a social-policy-oriented centre based on the two universities. Yet with all these investigations over a number of years, there is still no clear answer to the central question which this book addresses: 'Why is it, both in the past and even more so at present, in Northern Ireland, that the Roman Catholic population fares so much worse, in employment terms, than the rest of the population?'

Structure of the Report

The answer quoted at the beginning of the Introduction will not stand up to examination. Nor, we believe, is the phenomenon solely or even mainly due to discrimination by Protestant (or public) employers. Successive reports commissioned by the Fair Employment Agency have shown that in almost every part of the labour market studied, public and private sector alike, there is evidence of Roman Catholics doing less well than Protestants. In different branches of industry and services, this may take the form of failure to recruit any Catholic workers at all; in other cases there is an excess of Catholics at the lower end of the occupational scale, and a scarcity of Catholics in the supervisory, managerial and professional levels. How much of this is due to discrimination, how much to the level of qualifications of potential entrants, how much to the age and sex structure of entry over the years, is not the subject of the present study. We shall, instead, show what the position was in the early 1980s and demonstrate that these differences are substantial, whatever reasons we shall be able to find for this state of affairs in the concluding chapter.

In this report, we shall treat discrimination as a residual variable. That is to say we shall first examine, one by one, local labour market factors (mainly population structure on the supply side, and industrial structure on the demand side) and show the imbalances which arise. This will enable us to draw up a national labour market balance sheet which will show how much overall unemployment is structural – and that will give us some idea of what the position would be if population and job opportunities were equally distributed. We then return to the localities (and subregions) and we see how they differ from the national picture. We look at the qualifications of the labour force, and especially the entrants, and we see how far it could be true that those entering the labour markets in some areas appear not to be qualified for the opportunities which do arise, or may do in future.

We then return to the share of the Roman Catholics in the total labour force and various subsections of the occupational structure, and in industries, and we shall point out where they are seriously underrepresented in those sections of the labour market where prospects are relatively good. It is this final element which will lead us to suspect that discrimination remains a major part of the total picture.

There is, in Northern Ireland, no disadvantaged ethnic minority; but there is the same bias against women as there is elsewhere in the UK. Therefore Catholic women are doubly disadvantaged, and we shall argue that, because the country has consigned these women to the bottom of the labour market heap, relatively lower incomes in Roman Catholic households are bound to prevail, even without high male unemployment.

In another chapter, we shall briefly examine obstacles to mobility between areas of very high unemployment and those with relatively more favourable prospects. We will try to see whether there is evidence that the availability of housing, or the ownership of cars, affects the situation. These matters relate to physical constraints on mobility. We shall, in the end, be left with a question we cannot answer: whether there may not also be psychological barriers which make unemployed people stay where they are, rather than migrate to areas of greater prosperity. This is not something we can measure: we must, however, keep an open mind about the possibility that such obstacles do exist. If they do, they may be due to real fears of dangers involved in changing one's environment. If that is so, then none of the conventional measures to provide additional employment may have much effect on the perceived differentials.

We shall conclude with a review of the employment prospects of the Province within a general UK picture of long-term stagnation, without contributing any original analysis of Northern Ireland's industrial prospects. But we shall look at the scope for governmental measures of all kinds and try to assess the chances, especially for young people, in the Northern Irish economy overall and in the most disadvantaged districts in particular.

PART I

POPULATION AND LABOUR SUPPLY

1

Development of the Population of Northern Ireland

The population of Northern Ireland stagnated at about 1,300,000 from 1871 until after 1945; sizeable natural increases were cancelled out by emigration. It was a new phenomenon, after the last war, when emigration began to fall to smaller proportions, so that a continued high rate of natural increase resulted in quite a fast rate of population growth (see Table 1.1). Between 1961 and 1971 the

Table 1.1 *Population change in Northern Ireland, 1951–81*

	Persons	Percentage change
1951–61	+54,121	+3.9
1961–71	+111,023	+7.8
1971–81 (enumerated)	−54,106	−3.5
1971–81 (RG's revision)	+26,092	+1.7
1971–81 (PPRU estimate)	−3,825	−0.25

Sources: Own calculations based on Northern Ireland Census 1981, *Summary Report*, revised by Registrar General; last line is PPRU estimate SCG(86)2 (PPRU, 1986)

population grew by over 110,000 because an unprecedented natural increase of nearly 175,000 more than cancelled out a net outflow of 65,000 (General Register Office, 1984). The increase of 7.8 percent has to be compared with an increase of only 5.8 percent for the rest of the UK, despite a relatively much smaller rate of net out-migration (4.8 percent from Northern Ireland, compared with 0.5 percent from Great Britain). The fact that unemployment in 1971 was still under 38,000, or 7.4 percent of the labour force, suggests that the country had, in the previous decade, absorbed most of the young workers entering the labour market. If not, the out-migration figures would have been much higher, given continued boom conditions in Great Britain and a sharp reduction of overseas immigration after 1961. Clearly, however, the high rate of natural increase during the 1960s (12.2 percent compared with under 6 percent for Great Britain) would in itself be a potential cause for concern in the future unless the rate of increase in the number of jobs in the Province accelerated, or the rate of out-migration were to resume its former proportions.

The picture changes drastically when we look at the 1970s, for both Northern Ireland and the rest of the UK. The rate of increase of the Northern Ireland population slowed down to 1.8 percent for the decade. (If we accept the higher estimates of out-migration proposed by the PPRU that increase disappears altogether: PPRU, 1986). It was only 0.8 percent for Great Britain. Natural increase fell to 7 percent in Ulster, and to just over 1 percent in the rest of the UK. Emigration now amounted to at least 82,500 or 5.4 percent, and this removed a much larger slice of the natural increase than had been the case in the previous decade (77 percent of the increase lost by migration compared with 41 percent in the previous decade). Again, if one accepts the higher out-migration estimates recently published, this figure rises to 100 percent (PPRU, 1986). The rest of the UK lost only 215,000 net by migration, a negligible proportion. This migration, even in the much less prosperous 1970s, reduced the labour force potential in Northern Ireland quite noticeably, though as we shall see this occurred mainly in the earlier part of the decade. By 1983 the out-movement was a mere trickle compared with earlier periods, despite mounting unemployment. This is a clear reflection of the worsening prospects in the traditional outlets for surplus Irish labour, notably Great Britain.

These fairly violent swings in the recent evolution of the Northern Irish population are even more evident when one looks at the changes and present structure of the districts and regions. Taking only one simple measure, the sex ratio (females per thousand males), this shows very wide variations overall. There is a large excess of women in Greater Belfast (with its good employment opportunities for women in retailing, services generally, and administration). In contrast, there is a large excess of men in some of the predominantly rural areas of the west. From the point of view of a labour market analysis, one of the most crucial questions is the sex balance in the middle age groups, because it is there that we can most readily identify the effects of past out-migration (Tables 1.2 and 1.3; Figure 5). Thus, for England and Wales, the 45–54 age group is normally of equal size for males and females, the original male excess at birth having been evened out by higher mortality. In Northern Ireland this age group contains 8.4 percent more women. As we shall see, the relatively small size of the remaining male pre-retirement population affects employment prospects.

The age structure of the country also shows a large number of anomalies. Over 36 percent of the population, on the latest available estimates, are under 20; and 20 percent between 40 and 59. In Great Britain at the same time (1981) the younger age group contributed 27 percent of the total population, and the middle age group something over 23 percent (Figure 6). We shall demonstrate the effect of these overall differences on the labour market at a later stage. Here it is enough to say that Northern Ireland has always had an in-built momentum for growth, that this has in the past been modified by heavy out-migration, and that in the most recent periods the trend has once again increased the potential for imbalance.

It is quite clear, however, that this imbalance of age structure is due to past out-migration of young adults as well as to the relatively high fertility of those who remained behind. We cannot say what would have been the proportions between young and middle age groups in 1981 if there had not been such heavy out-migration between 1951 and 1961 and down to about 1966, because though undoubtedly there would now be more adults, they would also have married and produced children, many of whom would have to be added to the 1981 population under 20. There is little point in putting forward hypotheses concerning the relative fertility of out-migrants and others. What does matter is that the combination of heavy out-migration, especially of males in the boom years, and the reproductive performance of the remaining population, are equally important constituent parts of the unfavourable age structure now existing in Northern Ireland, in the face of economic decline.

Of the total population, the majority live in Belfast and the adjoining districts. (About 700,000 live in our area group I, which is essentially Greater Belfast, and at least 100,000 more in districts from which sizeable parts of the working population travel to work in Greater Belfast.) Although it is not true that this eastern part of the Province is 'Protestant', and the remainder 'Catholic', as is sometimes assumed, the fact is that the proportion of Catholics in the total population increases fairly steadily as one examines districts moving west and south from Belfast (Table 1.4, Figure 7). The effect of Partition was to exclude from the new Province those old Ulster counties which were almost exclusively Catholic (Donegal, Monaghan, Cavan), so that by definition all districts in the jurisdiction are now mixed, with only those in the extreme west and south having a Catholic majority.

It is also noticeable that the density of population declines sharply

Table 1.2 *Age distribution of population in Northern Ireland, 1966, 1971, 1981 (thousands)*

	Census 1966				Census 1971				Census 1981			
Age groups	Total	Males	Females	Percentage in each age group	Total	Males	Females	Percentage in each age group	Total	Males	Females	Percentage in each age group
0–4	160.5	83.0	77.5	10.8	156.2	80.3	75.9	10.2	130.9	66.9	64.0	8.4
5–9	146.3	75.1	71.2	9.9	157.1	81.1	76.0	10.2	134.3	69.2	65.1	8.6
10–14	132.0	67.7	64.3	8.9	143.6	73.7	69.9	9.3	148.6	76.2	72.4	9.5
15–19	125.7	63.5	62.2	8.5	126.3	65.1	61.2	8.2	149.3	77.3	72.0	9.6
20–24	106.4	52.4	54.0	7.2	114.9	59.3	55.6	7.5	125.5	64.7	60.8	8.0
25–29	87.8	43.4	44.3	5.9	101.9	51.6	50.3	6.6	102.7	52.7	50.0	6.6
30–34	83.8	41.1	42.7	5.7	86.7	43.5	43.2	5.7	101.3	51.1	50.2	6.5
35–39	84.8	40.9	43.9	5.7	82.4	40.7	41.7	5.4	94.6	47.2	47.4	6.0
40–44	88.2	42.7	45.5	5.9	84.3	40.7	43.6	5.5	82.9	40.8	42.1	5.3
45–49	83.6	40.9	42.7	5.6	86.1	41.7	44.4	5.6	77.6	37.8	39.8	5.0
50–54	83.4	40.2	43.2	5.6	80.1	39.1	41.0	5.2	78.8	37.6	41.2	5.0
55–59	77.5	37.0	40.5	5.2	78.5	37.4	41.1	5.1	78.1	36.9	41.2	5.0
60–64	68.6	31.3	37.3	4.6	72.0	33.1	38.9	4.7	70.8	32.7	38.1	4.5
65–69	58.2	25.3	32.9	3.9	60.2	26.4	33.8	3.9	66.0	29.3	36.7	4.2
70 and over	98.0	39.4	58.6	6.6	105.8	41.0	64.8	6.9	120.7	44.9	75.8	7.7
Totals	1,484.8	723.9	760.9	100.0	1,536.1	754.7	781.4	100.0	1,562.2	765.2	797.0	100.0

Source: Northern Ireland *Annual Abstract of Statistics* no. 3, Table 1.2

Table 1.3 *Age and sex structure of population of Northern Ireland by area group, 1981*

Area group[1]	Under 20	%[2]	20–39	%	40–59	%	60+	%	Total
Belfast									700,547
Males	118,058	35.2	95,657	28.5	71,795	21.4	49,602	14.8	335,112
Females	113,413	31.0	95,581	26.2	80,052	21.9	76,389	20.9	365,435
Northern									247,355
Males	46,981	38.3	34,933	28.5	24,227	19.8	16,470	13.4	122,611
Females	43,882	35.2	33,455	26.8	25,097	20.1	22,310	17.9	124,744
Southern									359,382
Males	71,329	39.8	49,698	27.8	34,227	19.1	23,779	13.3	179,033
Females	66,323	36.8	46,762	25.9	35,746	19.8	31,518	17.5	180,349
Western									254,873
Males	53,201	41.4	35,373	27.5	22,925	17.8	16,959	13.2	128,458
Females	49,966	39.5	32,671	25.8	23,394	18.5	20,384	16.1	126.415
Northern Ireland									1,562,157
Males	289,569	37.8	215,661	28.2	153,174	20.0	106,810	14.0	765,214
Females	273,584	34.3	208,469	26.2	164,289	20.6	150,601	18.9	796,943

[1] For definition of area groups see Appendix A.

[2] Percentages may not add up to 100 percent due to rounding.

Source: *Northern Ireland Annual Report 1981* (General Register Office, 1984)

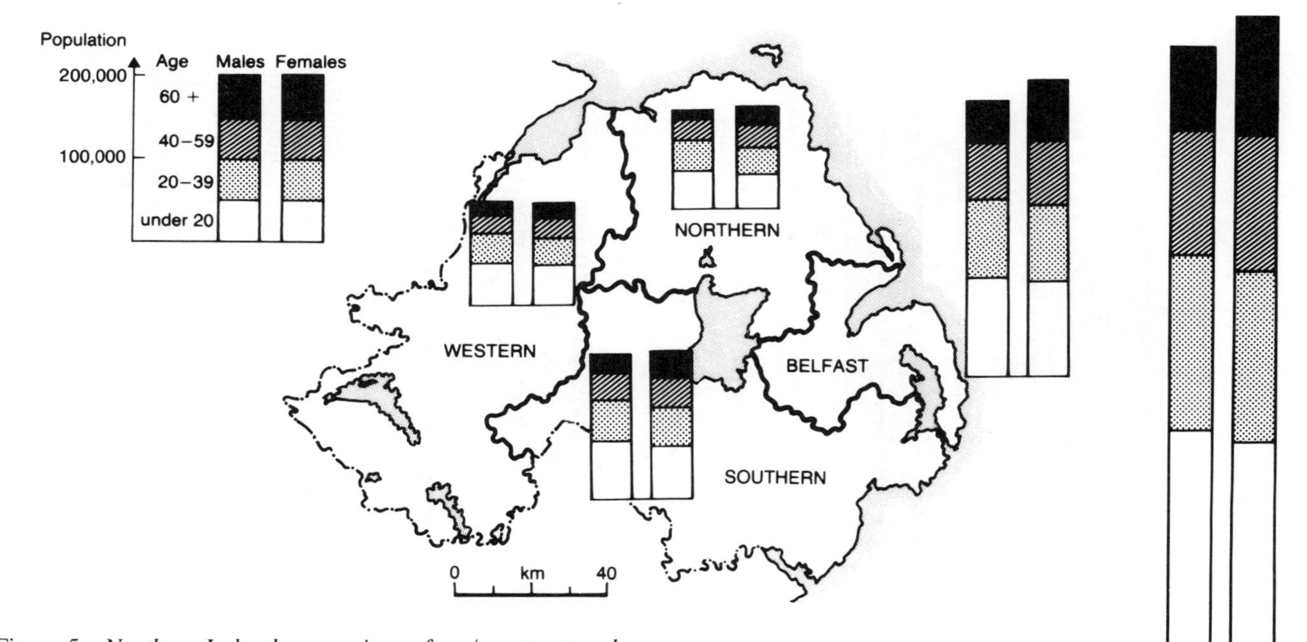

Figure 5 *Northern Ireland comparison of age/sex structure by area group, 1981 (see also Table 1.3)*

Source: Own calculations based on *Northern Ireland Annual Report 1981* (General Register Office, 1984)

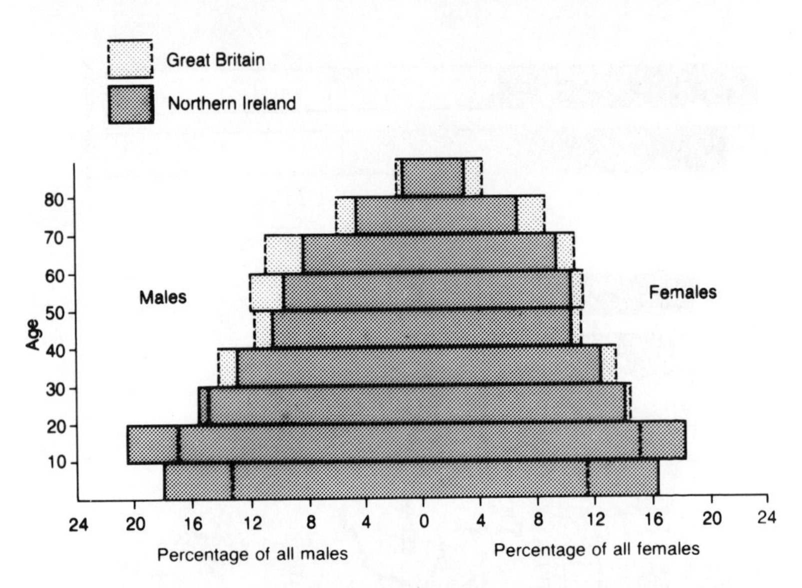

Figure 6 *Age pyramid showing differences in age structure between Great Britain and Northern Ireland, 1981*

Source: Office of Population Censuses and Surveys and Registrar General Scotland (1983); Northern Ireland Census 1981, *Summary Report*

as one moves away from Belfast; the country is predominantly rural, and in many parts of great scenic beauty.

The age structure of the districts varies considerably. When consulting the tables, it must be remembered that the smaller the area, and the more disaggregated the totals, the greater the chance of error. Clearly the whole population is young, but in the deprived areas particularly so. The most significant aspects, however, are the sex imbalances compared with the rest of the UK, and the differential proportion of the older people. A detailed examination of the age pyramids shows that the differences between districts are very large again, reflecting not so much differential fertility as different patterns of out-migration during the last fifty years.

As our Tables 1.3 and 1.4 and Figure 8 show, the overall anomalies are exacerbated at the district and the area group levels. This is not a phenomenon peculiar to Northern Ireland, but it does mean that unless there is a good deal of mobility within the labour market, local imbalances will be permanently reflected in differential economic activity rates and, at a time of recession, in unemployment.

Such differences also exist in Great Britain, but they are mostly caused by concentrations of retired persons in certain favoured

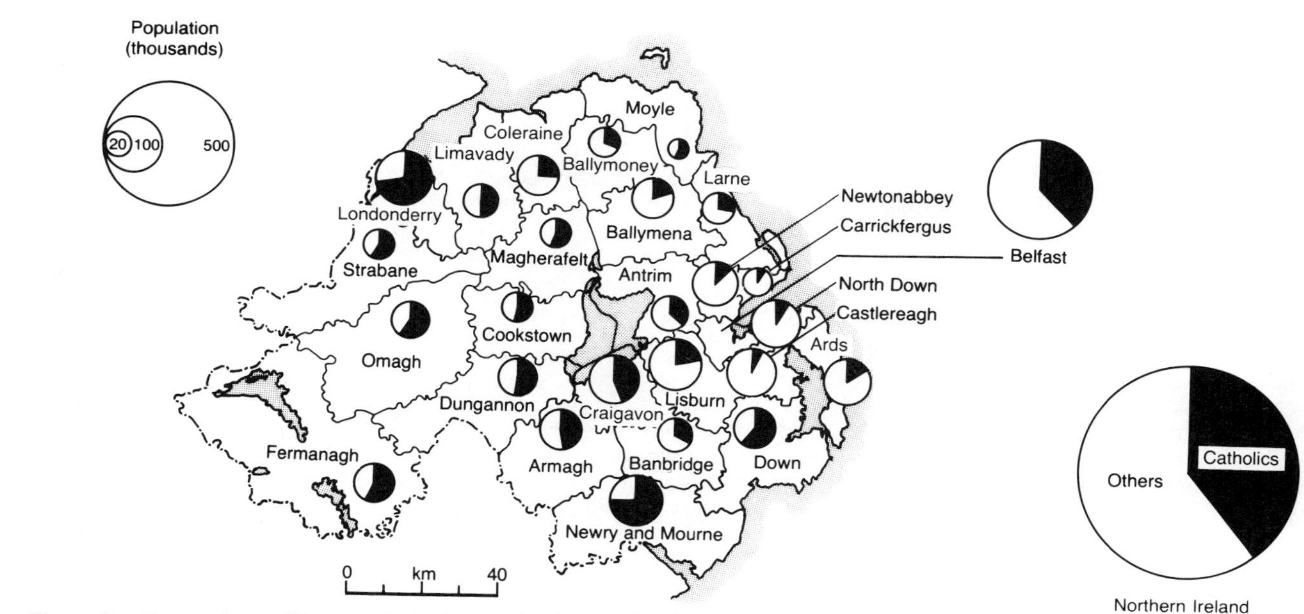

Figure 7 *Proportions of Roman Catholics in Northern Ireland by district, 1981 (see also Table 1.4). Area of circles is proportional to whole population, see scale*

Source: Northern Ireland Census 1981, *Religion Report*

Table 1.4 *Proportions of Roman Catholics in Northern Ireland in two age groups by area group, 1981*

	Under 15				15+ years				Totals			
Area group[1]	Catholics	Others	Total	%	Catholics	Others	Total	%	Catholics	Others	Total	%
Group I												
Ards	2,500	12,100	14,600	17	5,900	37,100	43,000	14	8,400	49,200	57,600	14
Belfast	36,900	38,400	75,300	49	88,000	166,700	254,700	35	124,900	205,100	330,000	38
Carrickfergus	700	6,600	7,300	10	2,000	19,100	21,100	9	2,700	25,700	28,400	9
Castlereagh	1,100	12,000	13,100	8	3,000	45,200	48,200	6	4,100	57,200	61,300	7
Lisburn	6,200	16,000	22,200	28	12,500	50,500	63,000	20	18,700	66,500	85,200	22
Newtownabbey	2,500	16,000	18,500	14	5,900	47,800	53,700	11	8,400	63,800	72,200	12
North Down	1,750	13,800	15,550	11	4,500	45,800	50,300	9	6,250	59,600	65,850	9
Subtotals	51,650	114,900	166,550	31	121,800	412,200	534,000	23	173,450	527,100	700,550	25
Group II												
Antrim	5,100	8,000	13,100	39	10,500	21,800	32,300	32	15,600	29,800	45,400	34
Ballymena	3,600	10,700	14,300	25	7,800	32,600	40,400	19	11,400	43,300	54,700	21
Ballymoney	2,300	4,000	6,300	36	4,500	12,100	16,600	27	6,800	16,100	22,900	30
Coleraine	3,500	8,700	12,200	29	8,400	26,400	34,800	24	11,900	35,200	47,100	25
Larne	2,200	5,100	7,300	30	5,600	16,500	22,100	25	7,800	21,500	29,300	27
Magherafelt	6,300	3,800	10,100	62	12,400	11,100	23,500	53	18,700	15,000	33,700	55
Moyle	2,400	1,600	4,000	60	5,500	4,900	10,400	53	7,900	6,400	14,300	55
Subtotals	25,400	41,900	67,300	38	54,700	125,400	180,100	31	80,100	167,300	247,400	32

Table 1.4 *continued*

Area group	Under 15				15+ years				Totals			
	Catholics	Others	Total	%	Catholics	Others	Total	%	Catholics	Others	Total	%
Group III												
Armagh	7,400	6,500	13,900	53	16,400	19,000	35,500	46	23,800	25,500	49,300	48
Banbridge	2,700	4,900	7,600	36	7,000	15,300	22,300	31	9,800	20,100	29,900	33
Cookstown	5,300	3,700	9,000	59	9,800	10,600	20,400	48	15,100	14,300	29,400	51
Craigavon	9,400	10,700	20,100	47	23,200	29,600	52,800	44	32,600	40,200	72,900	45
Down	10,300	4,600	14,900	69	20,500	17,700	38,200	54	30,800	22,300	53,100	58
Dungannon	7,600	5,600	13,200	58	15,800	16,400	32,300	49	23,500	22,100	45,600	52
Newry and Mourne	19,500	4,600	24,100	81	39,800	15,300	55,100	72	59,300	19,900	79,200	75
Subtotals	62,200	40,600	102,800	60	132,500	123,900	256,400	52	194,700	164,500	359,200	54
Group IV												
Fermanagh	8,600	5,400	14,000	61	20,900	17,000	37,900	55	29,500	22,400	51,900	57
Limavady	4,800	3,900	8,700	55	8,700	9,700	18,400	47	13,500	13,600	27,100	50
Londonderry	22,600	6,400	29,000	78	44,500	19,300	63,800	70	67,100	25,700	92,800	72
Omagh	10,000	4,000	14,000	72	20,800	11,500	32,300	64	30,800	15,500	46,300	66
Strabane	7,400	4,100	11,500	64	14,300	11,000	25,400	56	21,700	15,100	36,800	59
Subtotals	53,400	23,800	77,200	69	109,200	68,500	177,700	61	162,600	92,300	254,900	63
Northern Ireland	192,600	221,300	413,800	46.5	418,300	730,000	1,148,300	36.4	610,900	951,300	1,562,200	39.1

[1] For definition of area groups see Appendix A.

Source: Northern Ireland Census 1981, *Religion Report*, Table 3

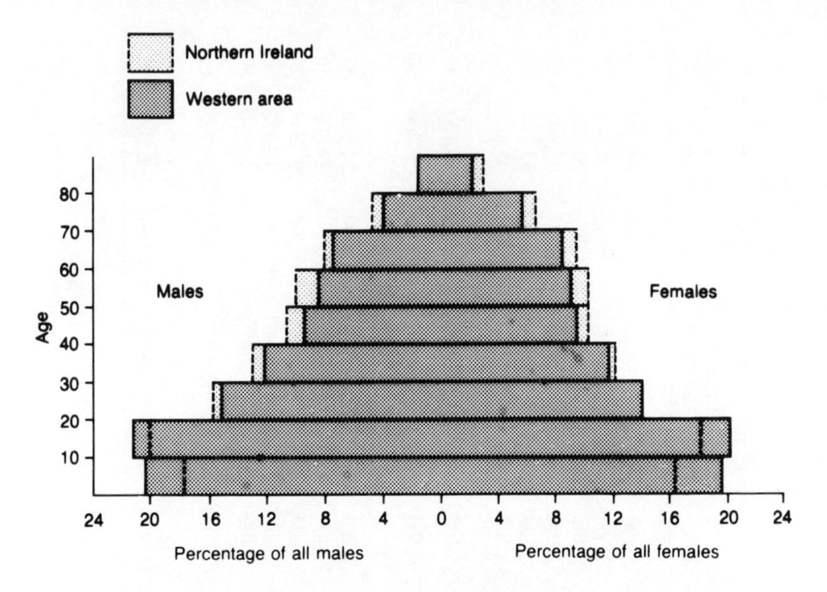

Figure 8 *Area group IV (Western) as proportion of population in decennial age groups compared with Northern Ireland structure, 1981*

Source: Northern Ireland Census 1981, *Summary Report*

coastal locations and, at the other end of the age range, by the young age structure of new towns and a few other special growth areas (for example, East Anglia). These differences are, however, small compared with those observed in the Northern Ireland subregions, and have little effect, at least so far, on the labour market. Given that Northern Ireland is the same size as five out of the eight English standard regions but has a much smaller population, one might expect local imbalances to be overcome more easily (than, say, between the Northern region of England and the South East). But that then raises the question of mobility within the regional labour market, and this will preoccupy us a good deal.

Looking into the future (see Table 1.5), if we follow the calculations of the Government Actuary, the problems caused by the age structure should get slightly easier for the rest of the century, as the number of potential labour market entrants drops quite considerably. This kind of projection does not take into account the possibility of a further drop in fertility, and also has to leave out the possibility of future sharp fluctuations in migration. After the turn of the century, on this projection, the imbalance between entrants and leavers would get worse again. But even when the relative proportions reach their potentially most favourable levels (in 1995) they would still be very much worse than that obtaining anywhere else in the UK. Only a combination of much higher economic activity rates (and lower unemployment rates) in the older working population, and a further reduction in fertility, could improve the outlook in the medium run. This will be discussed in detail in Part II.

Table 1.5 *Projected future population of Northern Ireland (thousands)*

Age	1981	1983	1985	1990	1995	2000	2005	2010	2015
0–4	132	139	143	156	159	153	147	150	157
5–9	131	123	125	138	151	155	149	143	146
10–14	148	141	133	123	136	149	153	146	140
15–19	151	151	146	129	119	131	145	148	142
20–24	132	138	143	139	122	112	124	138	141
25–29	104	112	121	137	133	116	106	119	132
30–34	97	94	96	116	132	129	111	102	114
35–39	95	97	96	93	113	129	125	108	99
40–44	81	84	90	93	91	111	127	123	106
45–49	77	78	78	87	90	87	107	123	119
50–54	77	77	75	76	82	86	85	103	118
55–59	76	74	73	70	71	78	81	79	97
60–64	68	71	70	67	64	65	71	74	72
65–69	61	58	57	61	58	55	56	62	66
70–74	52	53	53	47	50	49	46	48	52
75–79	35	36	38	39	35	37	37	35	36
80–84	21	21	21	22	24	22	24	23	22
85 and over	9	10	10	12	15	15	14	15	16
Total	1,547	1,557	1,568	1,605	1,645	1,679	1,708	1,739	1,775

Source: Northern Ireland *Annual Abstract of Statistics* no. 3, Table 1.7

As Table 1.4 and Figure 9 show, the position is worse within the Catholic community than for the population as a whole. We have not attempted a total reconstruction of the Catholic age structure based on our previous report (Eversley and Herr, 1985), which began with the school-age population, partly because of our inability to calculate detailed age-specific correction factors for the (underenumerated) adult Catholic population. However, it is clear that since Catholics constitute 46.5 percent of the population under 15, and only 36.4 percent of those over 15, the imbalance is worse within that group. Putting it another way, 31.5 percent of the Catholic population is under 15, but only 23.3 percent of the 'Protestants and others' group. Once again, this must be attributed not to high fertility alone, but also to the erosion of the older adult age groups by out-migration, differentially more so for Catholics.

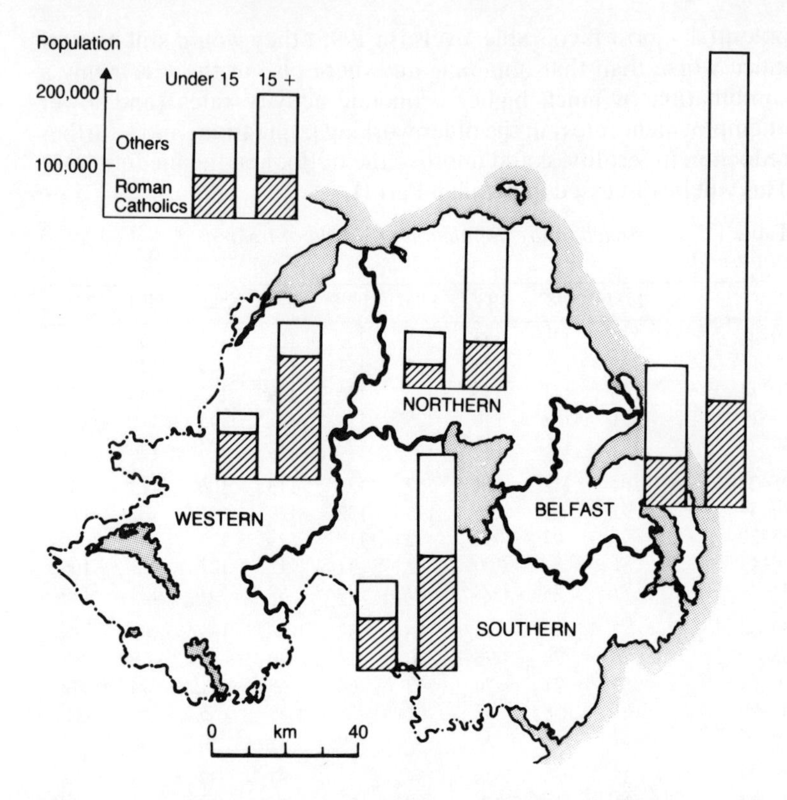

Figure 9 *Proportions of Roman Catholics in Northern Ireland in two age groups by area group, 1981*

Source: Northern Ireland Census 1981, *Religion Report*, Table 6

This can be demonstrated by looking at the religion tables of the 1971 and 1981 Censuses and observing the proportion of 'not stated' persons in the older age groups. This drops considerably for the middle-aged and older population, and thus intercensus comparisons can be used for evidence of religion-specific erosion of age groups by out-migration. Without entering here into the detailed calculations required, it is true to say that the reduction in the stated Catholic age groups concerned (40–49 in 1971, 50–59 in 1981) is far greater than that for the 'Protestant and other' denominations. It is, in fact, more than twice as high, and much greater than can be explained by the rise in the proportion of 'not stated' (who in any case have been shown to be only marginally more Catholic than non-Catholic), let alone by any potential differences in age-specific mortality. So we came to the conclusion that migration plays a large role in this difference. The difference in the sex ratio is also larger for Catholics. Even without pursuing the matter further back in history, then, the evidence of greater erosion by emigration is clear. (The other denominations are also affected by emigration, of course, because there too the reduction in older age groups is greater than can be explained by the life tables.)

At the local level this effect is again bound to be stronger, though it is quite impossible to calculate figures with any plausibility. Given the differences in underenumeration and in non-response to the religion question, and the vagaries of changes in the local settlement pattern, the calculation would be very difficult even without the changes in boundaries.

2
Migration

The extent of the reduction of the population of Northern Ireland by emigration has played a large role in the debates on the corrections required to convert the 1981 Census into a correct estimate. The same applies to the controversies about fertility, as regards both the differentials between the main religious groups, and the extent of the recent fall in rates for all groups.

Neither extent of migration, nor levels of fertility, enter directly into this report on labour market imbalances. However, there are some possible indirect consequences, especially when looking forward into the future, and it seems necessary to take a view on the matter. If it is a fact that out-migration was very much larger overall than we have assumed by our acceptance of the Registrar General's figures published in 1984 (General Register Office, 1984), then population in some districts from which there was particularly heavy out-migration might be rather smaller than we have assumed, with a possible consequent future easing of the employment problem.

The main contentious issue is whether net outflow from Northern Ireland was higher than the 82,000 estimated in 1984 by the Registrar General, or the rather higher figures published in successive estimates by the Planning Policy Research Unit (Morris et al., 1985). Much of this depends on the importance of estimates originating from the Irish Republic concerning migration from North to South (Garvey, 1985); and on one's acceptance of the estimates published in the International Passenger Survey, which forms the basis of estimates appearing under the aegis of the Office of Population Censuses and Surveys (OPCS) in Britain.

The limitations of the International Passenger Survey are quite clearly explained in the relevant publications of the OPCS (Office of Population Censuses and Surveys, 1975ff.). For illustrative purposes, the OPCS reports do allocate origins and destinations of migrants proportionately from the grossed-up sample in accordance with the responses to the questionnaire. However, it is also made perfectly clear in the 1983 OPCS *International Migration Report* that, because the sample is on average a little over 1 percent of all passengers, 'it is unwise to calculate standard errors for estimates of subsets of the population which are based on small samples.'

The estimate of the outflow to the Republic is derived partly from the Republic's Census in 1981, but mainly from the biennial Labour Force Survey. As regards the Census count (2,600 persons resident in the Republic in 1981 who a year earlier had lived in Northern Ireland: Garvey, 1985), this is partly offset by about 950 respondents in the Northern Ireland Census who declared their residence to have been in the Republic a year earlier, a net movement of about 1,500 (Northern Ireland Census 1981, *Migration Report*).

There are a number of other sources which have been investigated by those interested in the exact figure of migrants, such as the transfer of registrations in the National Health Service between Northern Ireland and Great Britain. An almost disproportionate amount of work has gone into ascertaining the grounds for believing that the Province lost 110,000 rather than 82,000 persons by migration during the last intercensal period (Compton, 1985a: 24).

In a sense, these debates are irrelevant here. Acceptance of the higher figure would reduce substantially, however, the number of persons originally stated as having been 'non-enumerated' – in fact the difference of about 30,000 between the Registrar General's 1984 figure and that adopted by PPRU in 1986. This difference itself would not affect many of the calculations which we shall undertake in this report. It would make a difference, however, if the additional out-migrants were very unequally distributed as between age and sex groups, or as between the districts from which they came. For this we have little evidence.

One aspect, however, of these new calculations is disturbing. In our earlier study (Eversley and Herr, 1985) we found 30,000 additional children by an analysis of school enrolment and child benefit figures. This figure would not have been too much out of line with the Registrar General's view of emigration, because these children would then account for 40 percent of the non-enumerated population (Morris et al., 1985). If, however, we accept the much higher outflow figure, the proportion of children amongst the non-enumerated would be much greater. The latest estimate (PPRU, 1986) gives a total underenumeration figure of only 44,500. Therefore the children we found would constitute 67 percent of the non-enumerated population. This seems absurd. It would mean that *all* the underenumerated households were large ones, consisting of two adults and four children (Morris et al., 1985). This would imply a serious bias, and it would mean that the population of the large (mainly Catholic) housing estates in and around Belfast accounts for most of the shortfall, with consequent implications for future labour supply in these areas.

All this may, however, be safely left to academic discussion as far as this report is concerned. Our analysis of imbalances in the labour market turns to a large extent on the ratio between the potential entrants to the labour market and the potential leavers. There is, in the further revisions of population that have been suggested, no mention of the older age groups, and the number of children is not seriously disputed. Therefore, even if the young and middle-aged groups really are somewhat smaller than we have assumed, this will not affect our analysis. What might be affected is the (Census) percentage of the unemployed, who would in all probability be heavily represented in the non-enumerated population: the downward revision postulated would reduce their numbers and therefore the percentage. Fortunately we do not rely very much on this statistic: our analysis is based on the insured population and the registered claimants. Equally, the question of occupational and industrial structure (of those employed at one stage or another in the period under discussion) is not affected by debates about the number of migrants. Our prognoses (and therefore policy conclusions) would not be changed by changing the 1971–81 migration figures: nobody has suggested that even if out-migration in the last decade was somewhat higher than has been assumed here, it could continue at that rate, given the deepening of the recession in the UK and the reduction in the acceptance of migrants by countries previously the destination of Northern Irish emigrants, and the sharp worsening of conditions in the Republic.

Age and Sex Structure of Migration

As has been pointed out, the exact size and composition of migration streams are hard to ascertain, except for the data derived from the transfer of NHS insurance cards between Northern Ireland and the rest of the UK. Tables 2.1 and 2.2 give two estimates of approximate age and sex composition; they are not, however, independent of each other as to sources of information. They are worth reproducing partly because they show how little accuracy is implied in the governmental statistics; sometimes when in- or out-movements are rounded off to the nearest 1,000, it is still not possible to state a 'net balance' of 1,000 when in- and out-movements are, say, 2,000 and 3,000.

A few conclusions, however, do emerge. As we would expect, migration is heavily concentrated on the 15–44 age group, and it can be assumed that this group consists to a large extent of married couples, given the number of children under 15 who are involved. Movement in the over 45 group is very limited, and almost non-

Table 2.1 *Estimated net migration from Northern Ireland, 1972–73 to 1982–83*

		Males				Females			
	Total	Under 15	15–44	45–64	65 and over	Under 15	15–44	45–64	65 and over
1972–73	12,484	1,100	3,722	572	161	1,337	4,395	810	387
1973–74	11,000	969	3,279	505	142	1,179	3,870	714	342
1974–75	16,000	1,410	4,770	734	206	1,713	5,630	1,039	498
1975–76	8,900	785	2,652	408	115	954	3,131	578	277
1976–77	8,200	723	2,445	375	106	878	2,885	533	255
1977–78	7,500	662	2,236	343	97	804	2,639	487	232
1978–79	5,700	1,033	1,855	195	75	445	2,065	10	22
1979–80	5,900	712	2,563	336	+53[1]	693	1,291	138	220
1980–81	6,200	318	2,630	176	19	748	1,933	98	278
1981–82	7,100	738	3,054	234	+121[1]	214	2,432	275	274
1982–83	5,300	811	2,236	—	133	549	1,380	191	—

[1] Net inflow.

Source: Northern Ireland *Annual Abstract of Statistics* no. 3

Table 2.2 *Internal UK population movements to and from Northern Ireland, 1981–84*[1]

	1981	1982	1983	1984
Totals				
In	7,000	7,000	7,000	7,000
Out	10,000	10,000	11,000	10,000
Net	−3,000	−3,000	−4,000	−3,000
Age 0–14				
In	—	2,000	2,000	2,000
Out	—	3,000	3,000	2,000
Net	—	−1,000	−1,000	—
Age 15–24				
In	—	2,000	2,000	2,000
Out	—	3,000	4,000	4,000
Net	—	−1,000	−3,000	−2,000
Age 25–44				
In	—	2,000	2,000	3,000
Out	—	3,000	3,000	3,000
Net	—	—	—	—
Age 45–64 (M)59(F)				
In	—	—	—	1,000
Out	—	1,000	1,000	1,000
Net	—	—	—	—

[1] The columns and rows may not add up due to rounding.

Source: Office of Population Censuses and Surveys *Monitors* MN 83/4, MN 84/4, MN 85/4

existent for the pensioner groups; there appears to be hardly any retirement migration in either direction (see Table 2.2).

As regards population dynamics, net movement undoubtedly means a proportionately large effect on the size of the future labour force, given the removal from the labour market of so many male wage earners and their children. We cannot be so sure about the female migrants. Until the late 1970s (Table 2.1) female emigrants exceeded males by a considerable margin. It must be presumed that those were the years when public services in Great Britain (notably the NHS) were still recruiting Irish girls. After 1979 there appears to have been a fairly dramatic turnround, with far fewer women migrating than men. Assuming that the proportion of married men moving out did not change drastically, we can only assume that the number of single women migrating decreased, and these might well form part of the increasing female unemployed labour force in Northern Ireland. (These assumptions are confirmed by inspection of the relevant tables of the Great Britain 1981 Census migration volumes: Office of Population Censuses and Surveys and Registrar General Scotland, 1983.) See also Table 2.2.

The net effect of such migration (which is, taking an overview of the decade, not very different from migration streams in earlier years: General Register Office, 1984) would normally be an ageing of the remaining population. Given, however, the continued relatively high fertility rates in the 1970s, this effect was not yet noticeable by the early 1980s. It may be assumed that such an effect must occur in the foreseeable future if net out-migration remains at its recent lower levels and fertility continues to fall.

Local Effects of Migration

In Tables 2.3 and 2.4 we show our estimates of the differential effect of migration on the population of the districts and area groups (see also Figure 10). Given the fact that we have little first-hand information on migration within the Province, the calculation presented here is based on the simple difference between the 1971 population plus natural increase, and the 1981 population calculated by correction from the Census, in the Registrar General's estimates. The sum of local surpluses (or deficits) arrived at in this way is much greater than the net migration balance with the rest of the world. Moreover, the gross streams in and out which lie behind these net figures are matters of pure guesswork, except for the year 1980–81, for which we have a summary of answers in the 1981 Census migration volume (Northern Ireland Census 1981, *Migration Report*): see Table 2.5. One is clearly aware that these figures refer to a population which

Table 2.3 *Migration balance 1971–81 by district, based on revised Census figures*

District	1971 population	Natural increase	1981 population	Migration residual	Total change (%)	Migration change (%)
Ards	46,778	3,699	57,600	+7,123	23.1	+15.2
Belfast	416,679	4,959	329,800	−91,838	−20.8	−22.0
Castlereagh	64,406	567	61,300	−3,673	−4.8	−5.7
Down	46,951	4,229	53,200	+2,020	13.3	+4.3
Lisburn	70,694	6,881	85,500	+7,925	20.9	+11.2
North Down	52,611	2,583	65,900	+10,706	25.2	+20.3
Antrim	33,998	6,177	45,600	+5,425	34.1	+15.9
Ballymena	48,998	4,121	54,800	+1,681	11.8	+3.4
Ballymoney	21,920	1,743	22,900	−763	4.5	−3.5
Carrickfergus	27,044	1,913	28,400	−557	5.0	−2.0
Coleraine	44,608	2,709	47,200	−117	5.8	−0.3
Cookstown	26,070	2,830	29,400	+500	12.8	+1.9
Larne	29,897	835	29,400	−1,332	−1.7	−4.4
Magherafelt	31,460	3,281	33,800	−941	7.4	−3.0
Moyle	13,979	954	14,400	−533	3.0	−3.8
Newtownabbey	66,915	6,693	72,300	−1,308	8.0	−1.9
Armagh	46,449	3,708	49,400	−757	6.3	−1.6
Banbridge	28,688	538	29,900	+674	4.2	+2.3
Craigavon	67,718	7,074	73,000	−1,792	7.8	−2.6
Dungannon	42,606	4,487	45,600	−1,493	7.0	−3.5
Newry and Mourne	72,368	8,613	79,400	−1,581	9.7	−2.2
Fermanagh	50,979	3,278	52,000	−2,257	2.0	−4.4
Limavady	23,809	3,346	27,100	−55	13.8	−0.2
Londonderry	84,901	14,037	93,200	−5,738	9.8	−6.7
Omagh	41,175	4,364	46,400	+861	12.7	+2.1
Strabane	34,364	3,935	36,900	−1,399	7.4	−4.1
Northern Ireland	1,536,065	107,554	1,564,400	−79,219	1.8	−5.1

Source: Own calculations based on Northern Ireland Census 1981, *Summary Report*, Table 2

overall has been seriously underenumerated, by as much as 3 or even 5 percent according to which estimate one takes. If, as has been postulated, this underenumeration is sharply concentrated on a small number of districts, then these migration figures may be misleading.

Evidence, however, is rather unreliable on this point. Comparing the revised estimates (PPRU, 1986) with the 1981 base figures in Table 2.5, we find that the differences are in fact largely accounted for by very few districts: of the 44,000 non-enumerated persons, over 30,000 can be allocated to Belfast, Londonderry, and Newry and Mourne, with only four other districts, all of them small, showing more than 5 percent difference between enumerated and estimated.

Table 2.4 *Migration balance 1971–81 by area group, based on revised Census figures*

Area group[1]	1971 population	Natural increase	1981 population (midyear)	Migration residual
I Belfast	745,000	29,000	701,000	−73,000
II Northern	225,000	20,000	248,000	+3,000
III Southern	331,000	32,000	359,000	−4,000
IV Western	235,000	28,000	255,000	−8,000
Northern Ireland	1,536,000	109,000	1,563,000	−82,000

[1] For definition of area groups see Appendix A.

Source: Own calculations based on Northern Ireland Census 1981, *Summary Report*, Table 2

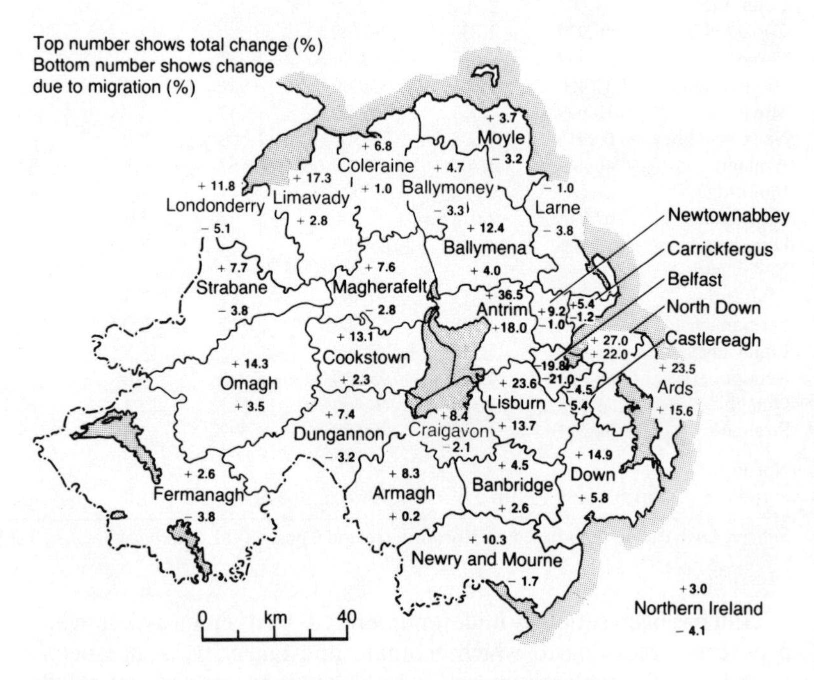

Figure 10 *Northern Ireland population change 1971–81 by district*

Sources: Northern Ireland Census 1981, *Summary Report; Northern Ireland Annual Report 1981* (General Register Office, 1984)

Therefore the general distribution of migration in the official Table 2.5 will imply an overestimation of out-migration for these seven districts. On the other hand Table 2.3, which is based on a ten-year balance and on the 1984 Registrar General's estimates of resident

Table 2.5 *Migrants within one year preceding Census day 1981, by district: numbers and proportions (per 1,000 resident population)*

District	Resident population at Census day	Migrants within the area		Immigrants from: All areas including abroad		Immigrants from: Elsewhere in NI only		Emigrants to areas in NI only		Migration balance (including immigrants from abroad)	
		Number	Proportion	Number	Proportion	Number	Proportion	Number	Proportion	Number	Proportion
Antrim	44,384	1,809	41	1,978	45	907	20	1,238	28	+740	+17
Ards	57,626	2,157	37	1,618	28	1,340	23	1,006	17	+612	+11
Armagh	47,618	1,401	29	893	19	649	14	650	14	+243	+5
Ballymena	54,426	2,736	50	1,228	23	786	14	859	16	+369	+7
Ballymoney	22,873	650	28	402	18	315	14	334	15	+68	+3
Banbridge	29,885	1,083	36	720	24	623	21	334	11	+386	+13
Belfast	295,223	14,680	50	5,701	19	3,984	13	6,937	23	−1,236	−4
Carrickfergus	28,458	974	34	805	28	643	23	573	20	+232	+8
Castlereagh	60,757	1,088	18	2,471	41	2,184	36	858	14	+1,613	+27
Coleraine	46,272	2,380	51	1,225	26	886	19	661	14	+564	+12
Cookstown	26,624	815	31	360	14	278	10	267	10	+93	+3
Craigavon	71,202	3,317	47	1,613	23	1,021	14	1,155	16	+458	+6
Down	52,869	2,413	46	2,130	40	901	17	828	16	+1,302	+25
Dungannon	41,073	1,472	36	477	12	380	9	575	14	−98	−2
Fermanagh	51,008	1,618	32	822	16	468	9	404	8	+418	+8
Larne	28,929	1,569	54	607	21	461	16	358	12	+249	+9
Limavady	26,270	1,337	51	1,109	42	370	14	187	7	+922	+35
Lisburn	82,091	2,861	35	3,710	45	2,133	26	1,241	15	+2,469	+30
Londonderry	83,384	3,819	46	1,567	19	461	6	842	10	+725	+9
Magherafelt	30,825	772	25	408	13	337	11	297	10	+111	+4
Moyle	14,252	374	26	296	21	240	17	216	15	+80	+6
Newry & Mourne	72,243	2,380	33	805	11	456	6	515	7	+290	+4
Newtownabbey	71,631	2,714	38	1,832	26	1,503	21	1,626	23	+206	+3
North Down	65,849	2,701	41	2,999	46	1,809	27	980	15	+2,019	+31
Omagh	41,159	1,190	29	1,180	29	372	9	552	13	+628	+15
Strabane	35,028	1,141	33	425	12	276	8	290	8	+135	+4
All districts	1,481,959	59,451	40	37,381	25	23,783	16	23,783	16	+13,598	+9

Source: Northern Ireland Census 1981, *Migration Report*, p. 4

population, implies an underestimation, especially for those areas where the correction as against the enumerated population was most drastic (again Belfast, Londonderry, and Newry and Mourne). Testing these various district assumptions against each other, we find that though absolute differences may be quite large, the distribution of migration flows hardly changes at all.

It would be very rash to multiply the figures in Table 2.5 by ten to arrive at an intercensal total, and arbitrary to average the figures from the various estimates of 1981 population. Unremarkably, however, the net annual balance published in the Census migration statistics has, in most cases, a plus or minus sign which is the same as in our Table 2.4, though the total magnitudes cannot be compared, because of the effect of net migration to other parts of the UK or abroad. There are a few exceptions to this. Thus Castlereagh lost population on balance after allowing for rather small natural increase figures, yet in 1980–81 it showed a healthy migration surplus. It takes little local knowledge to realize that this discrepancy is due to recent upswings in the private building sector in the area. Other districts with a much larger recent gain (at least from internal migration and movement from outside the Province) than might have been imagined from the decadal balance figures are Coleraine and Craigavon, as well as some of the western districts which, however, are in all probability also among the heaviest losers by external migration. Other areas show a 1981 gain from the Census, where the decadal figures show a loss.

On this evidence, we have mostly disregarded the Census and what follows is taken from our own computations based on the Registrar General's estimates (Figure 11). As Table 2.4 (area groups) shows, the loss remains most substantial in the Belfast area, which accounts for 90 percent of the net loss of the whole Province. Within that group, Belfast city shows a far greater loss (more than that of the whole country), whereas Ards, Lisburn and North Down show very substantial gains. Despite recent private sector building activity, Castlereagh, Carrickfergus and Newtownabbey have net population losses. Not too much should be made, perhaps, of the fact that the southern and eastern Belfast suburbs show large gains, and the northern ones losses: Antrim, large parts of which are in fact parts of the Belfast travel-to-work area, shows considerable gains. The fact remains, however, that the overall loss in the Belfast area group is far larger than can be explained by deconcentration of the population away from the city.

The other area groups do not show such massive changes. Both the Northern and the Southern groups were just about stable in the decade. Antrim and Ballymena contribute most of the Northern

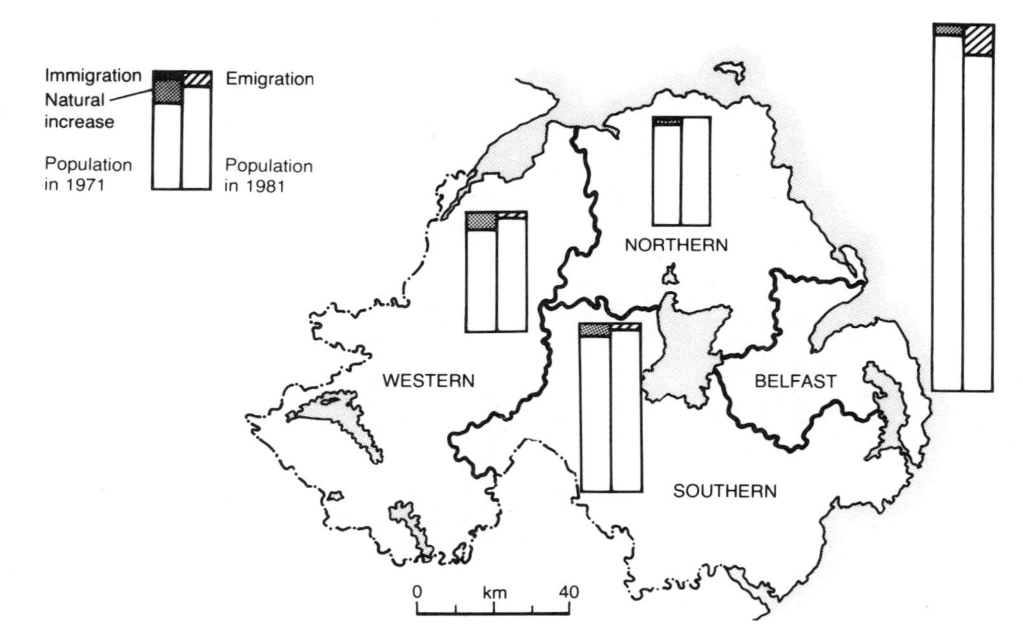

Figure 11 *Natural increase and migration balance 1971–81, Northern Ireland and area groups (see also Table 2.4)*

Source: Own calculations based on Northern Ireland Census 1981, *Summary Report*, Table 2

group's net gains, whereas Larne has a relatively heavy loss and other, more rural districts also experience moderately serious falls. In the Southern area, Down has quite large gains, and Banbridge and Cookstown smaller ones; all other districts have noticeable losses. The Western area group loses overall as we would expect, except for Omagh which records a small gain; Derry has absolutely and proportionately the heaviest losses.

Repeating the calculations on the basis of the larger net out-migration figure now assumed by some researchers, the picture does not change much. The figures for each group (as given in Table 2.4) become larger, because we have to find another 30,000 out-migrants, but the distribution is very similar: using our grouping, Greater Belfast loses just over 90,000 rather than 73,000; the Northern group loses 400 instead of gaining 3,000; the Southern group is down 8,500 instead of 4,000; and the Western group loses nearly 15,000 instead of 8,000. Thus only the Northern group, in the revised estimates, is shown to have had a very small net loss instead of a gain.

It will be seen from this comparison that, though magnitudes differ, this is a matter of degree rather than kind, and we shall leave the matter there.

How far these net balance gains and losses have altered the population composition of the districts can be estimated only by counting the difference between the 1971 population by age and sex group, and the 1981 corrected equivalents (ten years on) after allowing for age-specific mortality. We undertook this laborious exercise but it yielded few clear conclusions (however, see below for the older age groups), and it has also been undertaken by the PPRU for its revised estimates (PPRU, 1986). The areas of heaviest out-migration clearly lost people of working age, but since they were also areas of relatively high fertility, the overall effect on medium-term labour market prospects was not much different from that in other areas where the losses were smaller. On the other hand, the areas of heavy gross in-migration (that is, the Belfast suburban ring) clearly show a very young age structure (see Northern Ireland Census 1981, *Migration Report*: Table 2). Since, however, it may be assumed that the vast majority of their immigrants were in employment, and since their socio-economic composition is favourable, we should not be unduly disturbed by this phenomenon. The one exception is Lisburn, where very large Northern Ireland Housing Executive estates were built and where, given the socio-economic and religious structure of the working population and the number of children in these affected wards, we would expect considerable difficulties to arise in future (as indeed can be shown already by reference to the 1985 school-leaver unemployment situation: see Chapter 4).

Unfortunately we cannot tell just how much of the apparent erosion of some of the middle age groups is due to migration, except in aggregate. There would be no point in calculating the number of children born in each decade (even if we could reconstruct the post-1974 districts for earlier years than 1971), then deducting the local age-specific mortality ratios, and assuming that the rest of the difference is due to net out-migration. We have, however, performed this exercise for at any rate the major pre-retirement age groups by district for the last intercensal period, on the assumptions that we know the 1971 population fairly reliably, and that Northern Ireland age-specific mortality ratios would apply fairly equally throughout the Province, so that the resultant shortfall (or excess) in 1981 does represent true migration figures. We found that the main effect of the migration flows on local age structure has been that the number of potential retirers increased in relatively affluent owner-occupier areas, and decreased somewhat in the areas of high unemployment – but not sufficiently to affect the crucial ratios of entrants to leavers (see Chapter 4).

From this particular exercise, we would conclude that migration did not noticeably affect the age structure of the local labour market, though it clearly had an influence on overall size. Just when this migration occurred in the decade cannot be known for certain, but the evidence is that net out-migration fell considerably towards the end of the period, with the fall in employment opportunities in the UK.

We also used the tables published in the migration volume of the 1981 Census to test some general hypotheses about the extent of internal mobility within Northern Ireland (Northern Ireland Census 1981, *Migration Report*: Tables 3 and 4; for comparison see Great Britain Census, 1981, *National Migration Report*, Part I: Table 5). In general, it is supposed that mobility is lower compared with Great Britain, for a number of reasons. In fact we found that, in the year before the Census, about 7.4 percent of the resident enumerated Northern Irish population had moved, adding up movers between and within districts, and also figures for movement into Great Britain (10,000) and to the Republic and to other parts of the world. This gave us a total of nearly 110,000 movers – without, of course, telling us anything about how far people moved, or even if they crossed district boundaries.

In comparison, in the same year in Great Britain, 5 million people moved over shorter or longer distances, and that amounted to 9.7 percent of the resident population. On this evidence alone, we could not say that the people of Northern Ireland were significantly less likely to move than the British population. In Great Britain, too, the great majority of moves took place within administrative districts, or just across administrative boundaries.

The total gross movement out of Northern Ireland, at something under 1 percent of the whole population in one year, was more than double the percentage for the rest of the UK. Since that higher proportion, however, included all those who moved to Britain, and the whole of the out-movement from Britain related to migration to other countries, this part of the comparison is invalid. Purely internal movement, then, is a better indicator of the propensity to move. If we look at that parameter, the difference is more marked: in 1980–81, 6.5 percent of the population of Northern Ireland moved within the Province, and 9.3 percent of the population of Great Britain moved within that island. Given the fact that distances are overall much smaller in Northern Ireland, it would be true to speak of a lower propensity to move. If we pick out, say, the five districts comprising the Western group, and add up all those who moved to the Greater Belfast area (that is, movements from area group IV to area group I) we find only a few hundred such movers in that one year.

Out of all Northern Ireland one-year migrants, only 2.4 percent were retirement movers (men and women over 60 years of age moving to a different district). In Great Britain in the same year, of those moving within the island only, 10 percent were over 60, and 6.7 percent moved outside their former district of residence, which suggests that retirement migration as previously defined was much more important.

Could this account for the difference? If we take one-year movers (including migrants outside the country, gross) minus the over 60s, that is the working population and their dependants in Northern Ireland, we get 6.6 percent of the population moving house. For Great Britain, the figure is reduced to about 8.3 percent. Thus a significant difference still remains (and would do so even if we knew more about the non-enumerated population in Northern Ireland).

The conclusion therefore is that mobility within Northern Ireland seems somewhat lower than in Great Britain, but it is by no means negligible. The fact that a great deal (almost a third) of it consists of people moving out of Belfast is not in itself significant: probably the proportion of movers in Great Britain who at that time were leaving the large urban areas accounted for the same proportion.

It is interesting to check this evidence of recent mobility against the findings of the Continuous Household Survey, 1983 round (PPRU unpublished report, 1985). In answer to the question 'At the moment, are you seriously thinking of moving?', 13 percent of households answered in the affirmative. The proportions varied from 9 percent of owner-occupiers to 25 percent of private renters; 17

percent of NIHE or housing association renters also said 'Yes'. 'Seriously thinking' does not imply immediate action, but the answer is quite compatible with recently observed actual movement. More detailed breakdowns showed that 95 percent of intending movers wished to remain within Northern Ireland, a figure rising to 99 percent of those intending to move in the Western division (the CHS distinguishes between Belfast and other urban areas, and Eastern and Western non-urban areas). As for the reasons for wishing to move (a matter which we shall discuss again later), housing (size and quality) and environment figured much more heavily in the answers than anything else; neither 'the troubles' nor job prospects appeared to play much of a role. The same applies to the question analysing reasons for past moves. In passing, it may be noted that the total proportion of past movers discovered in the CHS is quite easy to reconcile with the 1980–81 proportion of migrants revealed by the one-year moving question in the Census, so that we can treat the CHS answers as a serious basis for the analysis of the prospects for the future.

Summary

Migration to the rest of the UK, and to other parts of the world, has always played an important role in checking the growth of the population of Ireland beyond the capacity of the country's agriculture and industry to employ additional labour. The scale of this migration diminished in recent decades, despite the continued differential in economic opportunities in Ireland and other countries. The much sharper increase in unemployment, especially in Northern Ireland, since the early 1970s, coincided with stagnation and decline elsewhere. The Irish Republic, which had attracted some return migration at the height of the boom years, afforded few opportunities after 1975. The rest of the UK saw a sharp rise in unemployment from the late 1970s, and though some migration from Northern Ireland continued, it was on a much smaller scale – not enough in fact to outweigh the increase in population by natural growth. The most recent information is that the Province continues to have a net migration deficit against Great Britain; movement is on a much reduced scale, and is not expected to increase.

As for the impact of migration on the balance of the religious denominations – that cannot be determined accurately. Traditionally more Catholics than Protestants emigrated, especially from the agricultural sector, and this was a factor in keeping the proportion of Protestants to Catholics more or less at the same level, despite higher Catholic fertility. Agriculture has now ceased to shed labour, and the construction industry, once one of the main customers for Irish labour in Great Britain, has been worse hit than other branches of production by the recession. This reduction in Catholic out-migration will probably lead to a change in the age balance within the minority population. On the other hand, it will make that much more difficult the task of providing jobs for people who in former decades would have emigrated, unless the Northern Irish economy grows more rapidly than in the past, and unless the difficulties of the Catholics in obtaining employment are considerably lessened.

3
Fertility

In many ways, the debates about the movement of fertility rates in Northern Ireland is peripheral to the main investigation which we have undertaken. The difference between current Roman Catholic fertility and that of other denominations, however, is clearly germane: if excess Catholic unemployment is blamed on excess Catholic fertility, then we clearly need to know whether this allegation is in fact correct and, if it is, just how large may be the difference and what are the time trends.

On this matter of absolute differences in recent years, there is no agreement among researchers. Estimates vary between 50 percent 'excess fertility' down to 36 percent at the most recent date (PPRU *Monitors*, no. 2/85; Compton et al., 1985). What is generally agreed is that fertility rates have fallen quite sharply in recent years, both for Catholics and for other denominations, though the extent of the fall is difficult to measure because neither the numerator nor the denominator used to compute birth rates can be accurately analysed by religious affiliation.

To a large extent, however, these debates concerning recent changes are not relevant here. The great majority of unemployed people in Northern Ireland were born before the decline in fertility set in. There is not much disagreement about the fact that at the time of the 1971 Census the differential was of the order of 54 percent, and though it is likely that the differential is now a good deal smaller, this is not going to affect the situation markedly until the end of the century.

There is now much harder evidence (Compton et al., 1985) on the differential, the decline in rates, and the decline in the differential, than was available when we did the groundwork for the present report. Nobody questions the very high rates of fertility up to the peak of the mid 1960s, or their gradual decline since then. The number of births has fallen quite sharply, whereas the women 'at risk' (that is, women aged 15–49) have increased (General Register Office, 1984). The number of women in the lower age groups will reach a plateau in the 1980s and thereafter decline quite rapidly. Neither marriages nor births show any signs, so far, of any change in the trends observed down to 1985. As in the rest of the UK, there is every possibility that marriage rates (expressed as persons marrying per 1,000 single in each age group) will rise when the delayed marriages take place; nobody, however, believes that these later marriages will then produce as many children as earlier ones would have done. Age-specific fertility rates have risen for the higher female age groups, but by nothing like as much as they have fallen for the younger groups. So the fall in fertility is continuing and may be expected to do so for some time yet.

Whatever happens, it cannot affect the analysis presented here. We print, for reference, some tables illustrating the main trends mentioned here, but no hard conclusions are drawn from them (see Tables 3.1–3.5). It has been suggested elsewhere that there is a

Table 3.1 *Age structure of Northern Ireland by area group in three age groups, 1981*

Area group[1]	0–14 years no.	%[2]	15–59(F)64(M) no.	%	Over 60(F)65(M) no.	%	Total population
I Belfast	166,682	23.8	423,422	60.4	110,443	15.8	700,547
II Northern	67,288	27.2	146,211	59.1	33,856	13.7	247,355
III Southern	102,769	28.6	208,447	58.0	48,166	13.4	359,382
IV Western	77,080	30.2	145,517	57.1	32,276	12.7	254,873
Northern Ireland	413,819	26.5	923,597	59.1	224,741	14.4	1,562,157

[1] For definition of area groups see Appendix A.

[2] Percentages may not always add up to 100 percent due to rounding.

Sources: Own calculations based on Northern Ireland Census 1971 and 1981, *Summary Reports*, as revised by the Registrar General

strong connection between rising female activity rates and a decline in fertility, but the nature of this connection has not been precisely analysed (Ermisch, 1982). Our Northern Ireland married women's economic participation rates are no help in the matter (see Chapter 7). It is probable that women's employment has been falling since it reached a peak in 1979, but this fall has not been associated with any rise in births. It would require an extremely detailed analysis of births to women by age, occupation and industry to be sure about what happens. In any case, a link between the changes in the employment market and reproductive patterns is so tenuous, even under conditions of relatively full employment, that we need not consider it further for Northern Ireland.

In Tables 3.3 and 3.4 we give the development of crude birth rates and total period fertility rates for Northern Ireland since 1979. It will be seen that, as in Great Britain, birth rates have oscillated around

Table 3.2 *Comparison of age distribution in Northern Ireland and United Kingdom, 1971 and 1981*

	Northern Ireland 1971		United Kingdom 1971		Northern Ireland 1981		United Kingdom 1981	
	no.	%[1]	'000s	%	no.	%	'000s	%
Age group 1								
0–14 male	234,617	15.4	6,873	12.4	212,287	13.6	5,876	10.7
0–14 female	221,352	14.6	6,515	11.7	201,532	12.9	5,579	10.1
Total	455,969	30.0	13,388	24.1	413,819	26.5	11,455	20.8
Age group 2								
15–64 male	439,625	28.9	17,176	31.1	478,787	30.6	17,710	32.1
15–59 female	419,865	27.6	17,545	31.6	444,810	28.5	17,778	32.3
Total	859,490	56.6	34,821	62.7	923,597	59.1	35,488	64.4
Age group 3								
Over 64 male	67,194	4.4	2,804	5.1	74,140	4.8	3,229	5.9
Over 60 female	136,987	9.0	4,502	8.1	150,601	9.6	4,940	9.0
Total	204,181	13.4	7,306	13.2	224,741	14.4	8,169	14.9

[1] Percentages may not always add up to 100 percent due to rounding.

Sources: Own calculations based on Northern Ireland Census 1971 and 1981 *Summary Reports*, as revised by the Registrar General.
For UK: *Annual Abstract of Statistics 1982* (Central Statistical Office, 1983)

the levels reached in the late 1970s, and there is not much change from what is a historically low level for an Irish population. Figure 12 shows the difference in birth rates by district in 1981.

The total period fertility rate has continued to fall since 1979. This is in contrast to England and Wales, where it rose from a low point in 1977 and by mid 1985 had again attained 1981 levels. So we are on fairly safe ground when we assume that fertility is still falling in Northern Ireland. Given the changed emigration figures discussed previously, with the sharp fall especially in the number of female migrants to Great Britain, births should be rising much more sharply than they have done: the fact that they do not indicates that either the Northern Irish women are not marrying, or (if they are) they are postponing child-bearing, or they are having smaller families, or all of these.

In the report on the religious composition of Northern Ireland previously cited (Eversley and Herr, 1985), we attempted some estimate of Catholic population in broad age groups. Basing ourselves solely on that exercise, we produced a table (Table 3.5) which shows the proportion of children in the age groups 0–14, to women aged 25–54, by religious affiliation. These calculations have been criticized on various grounds, and they are shown here not so much as a basis for making definitive judgements about relative fertility, as to show relative values at the local level. It will be seen that the differential in ratios within a Northern Ireland total of 56 percent (between 2.5 for Catholics and 1.6 for others) is not replicated evenly throughout the districts. In some cases (for example, suburban Belfast) the differentials are much smaller (Carrickfergus, Castlereagh, Newtownabbey, North Down), and in Belfast proper, as well as in some rural areas, they are larger. This suggests a fairly obvious trend: that where more of the resident population belong to the higher socio-occupational groups, the birth rate may well be lower than elsewhere, and if we could compute accurate local fertility ratios (to allow for the much younger age structure of these suburban populations) we should find them to be lower still. Within these lower reproductive tallies, the differences between religious groups are smallest, and where fertility is highest, the differentials are highest too. This last assertion in fact does not hold good for the Belfast figures shown. We have, however, previously demonstrated that Belfast suffers from a serious underenumeration problem, and so we should perhaps not attach too much significance to the fact that in that city the birth rate was apparently low, and the differential high.

In general, these ratios underline what has previously been said about present fertility trends: they are on the way down and, as they fall, so the differential between the religious groups also falls.

Table 3.3 *Evolution of live birth rates and total period fertility rates in Northern Ireland, 1979–84*

	Live birth rates	Total period fertility rate[1]
1979	18.3	2.65
1980	18.5	2.70
1981	17.5	2.46
1982	17.2	2.45
1983	17.3	2.44
1984	17.6[2]	2.44

[1] Total period fertility rate is the average number of children which would be born per woman if women experienced the age-specific fertility rates of the period in question throughout their child-bearing lifespan.
[2] First half only.

Source: Unpublished statistics communicated by the Northern Ireland General Register Office and the Office of Population Censuses and Surveys, London

It is not claimed that the children shown in the numerator of these ratios were in fact all born to women aged 25–54 at the date of the Census, nor that all the children ever born to the women in the denominator appear in the numerator. If there were important differences in the incidence of births by age of mother, this might invalidate these calculations as absolute indicators. However, as they are used here merely to illustrate further the large differentials in age structure which existed in 1981, and which have a bearing on the future of the labour market, we present them here with the reservations stated.

The outcome of these calculations is in fact much the same as that which we would have expected from the Registrar General's own allocation of births to social class in 1981 (General Register Office, 1984: 16–17) and from the results of the 1983 Fertility Survey (General Register Office, 1984). As in the case of migration, the analysis of fertility differentials is now largely of historical and academic interest; convergence has become so general that it may almost be taken for granted. The only point worth making, in advance of further analyses being published, is that there is some support for the inference that where Catholics have attained the economic status of their Protestant neighbours (notably in the Belfast region outer areas) their fertility is not so dissimilar. This does not prove that, if Catholics are given more chance of rising on the socio-economic status ladder, they will have smaller families; it might just as easily be said that only those who have smaller families to support experience higher geographical and social mobility. However, the converse is much more likely to be true: in the more rural areas where employment opportunities are few (except for self-employment, and the not

Table 3.4 *Live birth numbers and rates, by district, 1978–81*

District	Live births				Rate per 1,000 of the population			
	1978	1979	1980	1981	1978	1979	1980	1981
Ards	977	1,040	1,075	908	18.4	19.5	20.0	15.8
Belfast	4,768	5,321	5,322	4,920	13.5	15.2	15.4	14.9
Castlereagh	568	548	435	612	8.9	8.6	6.9	10.0
Down	981	1,015	1,055	1,042	20.1	20.8	21.4	19.6
Lisburn	1,329	1,419	1,524	1,423	16.1	17.0	18.1	16.6
North Down	924	1,095	1,098	1,018	15.0	17.5	17.4	15.5
Antrim	835	986	995	936	20.9	24.0	23.8	20.5
Ballymena	962	993	1,077	1,002	18.1	18.7	20.0	18.3
Ballymoney	408	406	508	381	18.5	18.2	22.6	16.6
Carrickfergus	446	445	431	444	16.0	16.2	15.2	15.6
Coleraine	675	788	760	734	14.8	17.2	16.5	15.6
Cookstown	520	565	453	525	18.6	20.0	16.0	17.9
Larne	397	394	403	409	14.0	13.9	14.2	13.9
Magherafelt	602	674	736	655	18.6	20.5	22.2	19.4
Moyle	271	254	192	251	20.8	19.7	15.1	17.4
Newtownabbey	1,257	1,256	1,330	1,267	16.8	16.5	17.3	17.5
Armagh	794	872	863	827	16.7	18.2	18.2	16.7
Banbridge	349	407	466	418	12.1	14.1	16.0	14.0
Craigavon	1,340	1,561	1,471	1,410	18.3	21.3	19.9	19.3
Dungannon	932	951	988	950	22.0	22.4	23.1	20.8
Newry and Mourne	1,649	1,777	1,756	1,742	21.7	23.1	22.8	21.9
Fermanagh	941	1,005	1,094	1,006	18.5	19.7	21.5	19.4
Limavady	481	517	486	532	18.8	20.6	19.3	19.6
Londonderry	2,196	2,172	2,376	2,235	25.0	24.3	26.1	24.0
Omagh	907	1,023	1,025	1,021	21.8	24.2	23.9	22.0
Strabane	730	684	663	634	20.5	19.1	18.4	17.2
Northern Ireland	26,239	28,178	28,582	27,302	17.1	18.3	18.5	17.5

Source: *Northern Ireland Annual Report 1981* (General Register Office, 1984)

very well paid public sector jobs), where housing is relatively cheap and community bonds strong, there is much less likelihood of lower fertility levels being reached so quickly. This at least needs to be taken into account when we are planning for future levels of employment.

Closer inspection of Table 3.5 shows the variation by religion in our rough ratios (from 1.3 for others and overall in Castlereagh, to 3.0 for Catholics in Cookstown and Moyle) to be greater than that for overall ratios (1.3 up to 2.4). When we look at the 1981 crude birth rate statistics in the last column of Table 3.4 or 3.5, this turns out to have only slightly less variance, from 13.9 in Larne to 24.0 in Derry. However, within the Catholic community, the variations are again almost as large: from 1.8 and 1.9 in Belfast suburban areas to the two districts with 3.0 already mentioned. For the 'others' the variation is

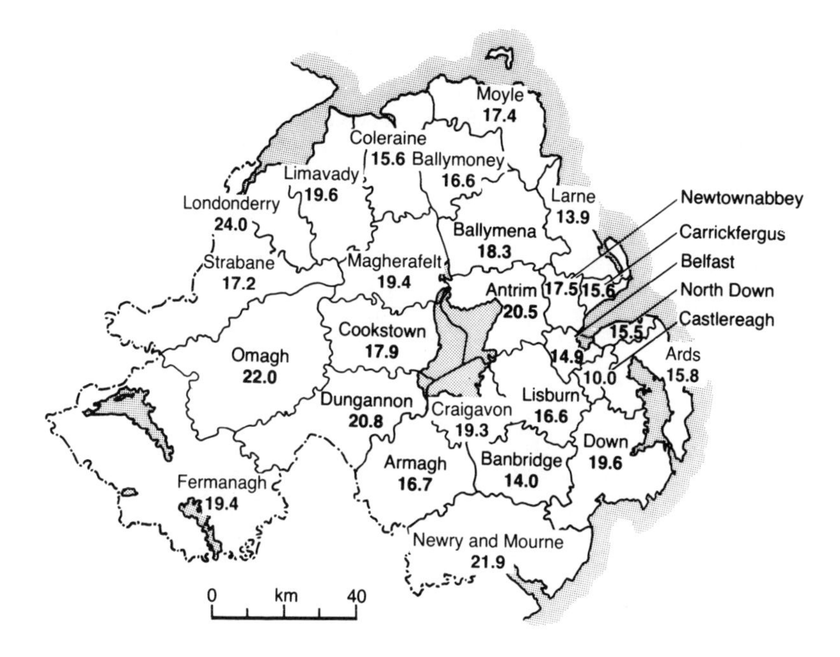

Figure 12 *Birth rates per 1,000 population by district, 1981 (see also Table 3.4)*

Source: Northern Ireland Annual Report 1981 (General Register Office, 1984)

much smaller: from 1.4 in Down and North Down to 1.9 in Limavady. For the Catholics, the highest ratio is two-thirds larger than the lowest; for the 'others' the highest is only just over one-third larger than the lowest.

It is not difficult from that to conclude that not only is the scope for reduction among the Catholics much larger, but it must be happening already, and it is primarily a question of class or socio-economic group, of tenure and probably of age structure. We can leave the exact evaluation of these striking differences, and the trends, in the hands of those analysing the 1983 Fertility Survey: we are not likely to come to any different conclusions.

The most important effect of fertility, as regards the objectives of this report, is to reinforce the conclusion that the future problem is likely to be much larger in some areas than in others. It is wrong, however, to single out past fertility as the *cause* of the gap: as we shall see, there are a great many differences between the districts and area groups other than in religion and fertility. It is the employment and industrial structure which will turn out to be central to the logical

Table 3.5 *Ratio of children aged 0–14 years to women aged 25–54 years, all denominations, Roman Catholics and others, 1981*

District	Children 0–14 to women 25–54			Live births per 1,000 of the population
	All	RC	Others	
Antrim	1.9	2.4	1.7	20.5
Ards	1.6	2.2	1.5	15.8
Armagh	2.0	2.5	1.7	16.7
Ballymena	1.7	2.3	1.6	18.3
Ballymoney	1.9	2.8	1.6	16.6
Banbridge	1.7	2.0	1.6	14.0
Belfast	1.8	2.5	1.4	14.9
Carrickfergus	1.6	1.8	1.6	15.6
Castlereagh	1.3	1.7	1.3	10.0
Coleraine	1.8	2.3	1.7	15.6
Cookstown	2.3	3.0	1.7	17.9
Craigavon	1.9	2.0	1.7	19.3
Down	2.1	2.6	1.4	19.6
Dungannon	2.2	2.7	1.8	20.8
Fermanagh	2.1	2.6	1.6	19.4
Larne	1.7	2.2	1.6	13.9
Limavady	2.4	2.9	1.9	19.6
Lisburn	1.7	2.2	1.5	16.6
Londonderry	2.4	2.7	1.7	24.0
Magherafelt	2.3	2.8	1.7	19.4
Moyle	2.2	3.0	1.6	17.4
Newry and Mourne	2.3	2.6	1.7	21.9
Newtownabbey	1.5	1.8	1.5	17.5
North Down	1.5	1.9	1.4	15.5
Omagh	2.4	2.8	1.7	22.0
Strabane	2.4	2.9	1.8	17.2
Northern Ireland	1.9	2.5	1.6	17.5

Sources: General Register Office, 1984; Eversley and Herr, 1985; Northern Ireland Census 1981, *Religion Report*

structure of the differentials. Fertility, in as far as it has contributed in the past, is clearly changing in such a direction as to make one more optimistic for the future; but the structure of the occupational categories, and especially of the industry divisions, gives much more cause for pessimism.

Summary

Catholic fertility has for long been higher than Protestant fertility, however measured. The outcome has been, at any one time, that there have been more new Catholic entrants to the labour force, than

Protestants, relative to the number of working adults in each population.

In recent years, fertility has fallen for both Catholics and Protestants. The differential, however, has remained, albeit at a much lower level of child-bearing, and hence the expected future entry of Catholic school-leavers into the labour market will be smaller in relation to the working population. Both Catholic and Protestant fertility levels, however, are still high compared with the rest of the United Kingdom, though a good deal lower than in the Irish Republic. As will be shown later, higher fertility is not the main cause of higher Catholic unemployment. The number of adult workers in each denomination depends on the size of the working cohorts. The Catholic men in particular have had a high propensity to migrate outwards in search of work; Catholic unemployment has been higher than among Protestants, and thus the competition for workplaces in a labour market divided by sectarian considerations has been more severe for Catholics, not only because of fertility. The same applies to local labour markets: the situation has been worse for the Catholic population in the peripheral areas, and not only because of the number of children.

One cannot forecast the future with certainty, but at least until the mid 1980s the decline in fertility has continued in Northern Ireland. In common with the rest of the UK, Northern Ireland has experienced a very slight rise in fertility again in the most recent years, after a steady fall since the low point of 1977, but even in 1985 the fertility of the Northern Ireland population was lower than it had been for the whole of the UK in 1964.

PART II

LABOUR DEMAND AND MARKET BALANCE

4

Development of the Labour Market

Introduction

The first part of this report examined in some detail the development of the present population of Northern Ireland, including some recent changes in migration patterns and fertility. We now need to summarize these details to begin our analysis of the problems of local labour markets.

Ulster, like the rest of Ireland, has been a country of strong out-migration for a long time. There have been fluctuations, to be sure; periods of relative prosperity in southern England in particular, even periods of labour shortage in the Republic, have slowed down the net outflow and on occasion reversed the movement. However, throughout the last 150 years the population pyramid has been a curious one, much broader at the base than that usual in most countries of Western Europe, and occasionally highly concave in its central layers. This has been due to two factors: first, the higher fertility of all groups of the Irish population, North and South, Catholic and Protestant; and secondly, the high propensity to out-migration shown by the young and young middle-aged groups, particularly single males and single women and sometimes married couples. Some of these out-migrants returned after retirement, and in any case not even the majority of working-age people migrated, so that at the top (the aged) the pyramid returns to a normal shape. However, the main characteristic of the labour supply structure of an area of high out-migration is that, whatever the fertility rates, there are relatively far more entrants to the labour market than retirers from it. There is therefore a permanent imbalance, or potential imbalance. In years when labour markets elsewhere can absorb the extra entrants, or when there is an expansion of the domestic demand for labour, this surplus can be managed. If the age of leaving full-time education rises, this also brings temporary relief.

This basic labour market imbalance is not confined to Ireland. It has been observed in many countries where fertility has been high, for a long time, compared with that obtaining in neighbouring or other accessible countries, and where economic growth rates, or at least demand for additional labour, have been low compared to other economies. It follows that excess unemployment occurs in such countries when for any reason out-migration is slowed down, unless there is an increase in the rate of economic growth of a kind that demands a great deal more labour. Adjustments in fertility can only help to moderate the effect of such changes in the long run (one thinks particularly of the countries from which the European 'guest workers' originated, and Mexico).

The degree to which the potential retiring group is actually still in work when it reaches pensionable age is also important. If a high proportion of the pre-retirement age group is unemployed, or economically inactive, or if a high proportion of them is self-employed, possibly in occupations where there is no natural succession between generations, then the imbalance between entrants and leavers becomes stronger. If, for instance, the number of agricultural holdings, and the number of small retail or artisanal establishments, is reduced through amalgamation, rationalization and mechanization, then self-employed retirers are not replaced.

To this we have to add the fact that although the retirers in the oldest age group are almost all men, the entrants are both boys and girls. It is true that women also leave the labour force, often at an early age because of child-bearing, or because of other domestic commitments when they become older, but the imbalance still exists.

This phenomenon is not confined to Northern Ireland. In all economies, the changeover to service-type employment, the growth of retailing, catering, banking and financial services, means more jobs for women and in many cases part-time workers. If in the past the economy was more heavily dependent on manufacturing, then the proportion of men in the age groups reaching retirement from the labour force will always be higher than the proportion of men finding employment at the lower end of the age distribution. Nor is it as certain as it once was that the women entrants will retire from the labour market at the onset of their reproductive lifespan. (That event itself occurs rather later now than it did in the past.) They are more likely to resume work within a relatively short time, at least on a part-time basis. However precarious the position of many young female entrants into the labour force (McCartney and Whyte, 1984), their sheer numbers effectively lower the employment prospects of the male entrants. This is no argument against the training and employment rights of women, but it needs to be considered in any overview of the labour market. Unfortunately this effect cannot be quantified, and thus we shall in this report mainly compare male entrant and leaver groups.

The next stage of the argument is that the changes in industrial structure in Northern Ireland have militated against the western areas and the locations of old staple industries (for example, textile and clothing), just as they have done in the rest of the UK, and that the main new service-type employment has been created in the eastern urban areas. Industrial investment, such as it was, was also heavily concentrated on the eastern areas (Harrison, 1980; Bradley et al., 1984; Northern Ireland Economic Council Report no. 23). The new jobs have often required a higher educational achievement than those which were destroyed. Women have benefited rather than men, in many industries. These are familiar phenomena in the whole UK labour market, where the new opportunities of the London region for many years almost outweighed the loss of opportunities elsewhere. Some of the new jobs are population related (for example retailing and professional services, especially public sector employment in health and education) but with a distinct bias again in favour of the urban eastern areas, characterized by high purchasing power and the concentrations of higher echelons in both public and private services near corporate headquarters. There is everywhere some interaction between the propensity to create new jobs in high- (or higher-) technology industries and services, and the presence of institutions of further and higher education. So the differential evolution of local labour markets will form a large part of the report.

Overview of the Labour Force

In Tables 3.1 and 3.2 we showed the generalized age structure of Northern Ireland, both by area groups and in relation to the UK, and the change in that age structure between 1971 and 1981. We first note that Northern Ireland has a high dependency ratio. This is measured by comparing the size of the potentially economically active age groups – men aged 15–64 and women aged 15–60 – with the under 15 age group and the pensioners (over 65 for men, over 60 for women). For the UK in 1981 that ratio was 181, or 1.8 potential workers for every dependant. In Northern Ireland it was 144, or 1.4 potential workers per dependant. In both cases this was an improvement against 1971: despite an increase in the relative size of the retirement group, there was a sharper reduction in the proportion of children. This means that, apart from everything else, to achieve the same standard of living for the population as a whole, earnings per person of working age should be higher in Northern Ireland than in the UK as a whole to achieve the same living standards.

We will not pursue this line of argument, because the existence of a national system of social security makes any calculation of actual earnings somewhat irrelevant. However we recognize (without quantifying the statement) that, given much higher unemployment, lower labour force participation rates and lower earnings, a much larger proportion of all household incomes must come from transfer incomes (pensions, allowances, benefits): this is amply confirmed by the Family Expenditure Surveys (Department of Employment, 1985: 91; Morris and Wilson-Davis, 1983). The additional costs arising from higher dependency ratios fall predominantly on central government, and thus it is not believed that a higher dependency ratio is of itself a cause of poverty. However, whatever the gross product of an economy, the more people have to share it, the lower the income per person; and there are a great many studies both in Great Britain and in Northern Ireland to show that families with three or more children are noticeably worse off than those with two children or less. This matter is, however, outside the scope of this investigation.

Within Northern Ireland, there are further significant variations in the dependency ratio: it is lowest in the Belfast group (area group I) and highest in the Western group (area group IV) (see Table 3.1). As it happens, area group I also has the highest proportion of older people (because it includes some preferred retirement areas) but the lowest proportion of children. The reverse situation applies in area group IV. As we shall see, the differences in proportion have important consequences for future labour market structures.

In Tables 4.1 and 4.2 we show the size of the potential labour market entrant groups in three quinquennial periods: 1986–90, 1991–95 and 1996–2001. These calculations are independent of future fertility: all those included in these tables were alive in 1981. However, the entrant groups could be changed by out-migration, normally if their parents emigrated in the meantime. This is true overall, but even more so at area group and at district levels. We read these tables therefore by saying that overall, even without out-migration, the entrant groups will be substantially smaller in the early 1990s, with some further improvement, that is decline, before the end of the century. As we would expect, the reduction will be proportionately sharpest in area group I, which will admit 13.7 percent fewer boys and girls in ten years' time, whereas the improvement will be minimal in the other three groups (see also Figure 13).

Table 4.1 *Potential labour market entrants 1986–2001, males, females and totals, by district*

	1986–90 (aged 10–14 in 1981)			1991–95 (aged 5–9 in 1981)			1996–2001 (aged 0–4 in 1981)		
District	Male	Female	Total	Male	Female	Total	Male	Female	Total
Antrim	2,316	2,151	4,467	2,221	2,192	4,413	2,181	2,048	4,229
Ards	2,618	2,482	5,100	2,575	2,278	4,853	2,372	2,304	4,676
Armagh	2,415	2,369	4,784	2,331	2,232	4,563	2,333	2,151	4,484
Ballymena	2,718	2,537	5,255	2,426	2,275	4,701	2,275	2,069	4,344
Ballymoney	1,184	1,114	2,298	1,053	946	1,999	991	923	1,914
Banbridge	1,449	1,296	2,745	1,255	1,202	2,457	1,227	1,202	2,429
Belfast	14,181	13,936	28,117	11,875	11,473	23,348	12,167	11,649	23,816
Carrickfergus	1,445	1,323	2,768	1,254	1,179	2,433	1,074	1,081	2,155
Castlereagh	2,780	2,681	5,461	2,128	1,925	4,053	1,831	1,772	3,603
Coleraine	2,367	2,376	4,743	2,015	1,972	3,987	1,801	1,722	3,523
Cookstown	1,491	1,478	2,969	1,535	1,396	2,931	1,558	1,500	3,058
Craigavon	3,718	3,581	7,299	3,365	3,139	6,504	3,189	3,067	6,256
Down	2,759	2,547	5,306	2,531	2,323	4,854	2,469	2,265	4,734
Dungannon	2,352	2,116	4,468	2,269	2,128	4,397	2,235	2,188	4,423
Fermanagh	2,449	2,332	4,781	2,449	2,265	4,714	2,284	2,216	4,500
Larne	1,547	1,389	2,936	1,234	1,164	2,398	959	992	1,951
Limavady	1,537	1,410	2,947	1,492	1,471	2,963	1,400	1,342	2,742
Lisburn	4,151	3,886	8,037	3,774	3,496	7,270	3,534	3,387	6,921
Londonderry	5,126	4,852	9,978	4,852	4,548	9,400	4,965	4,696	9,661
Magherafelt	1,861	1,666	3,527	1,703	1,619	3,322	1,683	1,598	3,281
Moyle	761	641	1,402	720	644	1,364	612	614	1,226
Newry and Mourne	4,300	4,189	8,489	4,071	3,698	7,769	4,033	3,817	7,850
Newtownabbey	3,420	3,199	6,619	3,089	2,964	6,053	2,964	2,853	5,817
North Down	2,809	2,651	5,460	2,633	2,475	5,108	2,531	2,483	5,014
Omagh	2,346	2,266	4,612	2,348	2,270	4,618	2,336	2,349	4,685
Strabane	2,062	1,952	4,014	2,049	1,801	3,850	1,866	1,749	3,615

Source: Own calculations based on *Northern Ireland Annual Report 1981* (General Register Office, 1984)

Table 4.2 *Potential labour market entrants 1986–2001, males and females, by area group*

Area group[1]	1986–90 (aged 10–14 in 1981)	1991–95 (aged 5–9 in 1981)	1996–2001 (aged 0–4 in 1981)
I Belfast	61,562	53,118	52,002
II Northern	24,636	22,184	20,468
III Southern	36,060	33,475	33,234
IV Western	26,332	25,545	25,203
Northern Ireland	148,590	134,322	130,907

[1] For definition of area groups see Appendix A.

Source: Own calculations based on *Northern Ireland Annual Report 1981* (General Register Office, 1984)

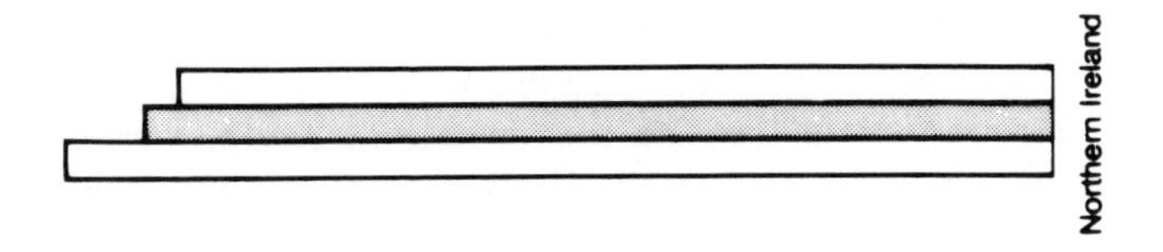

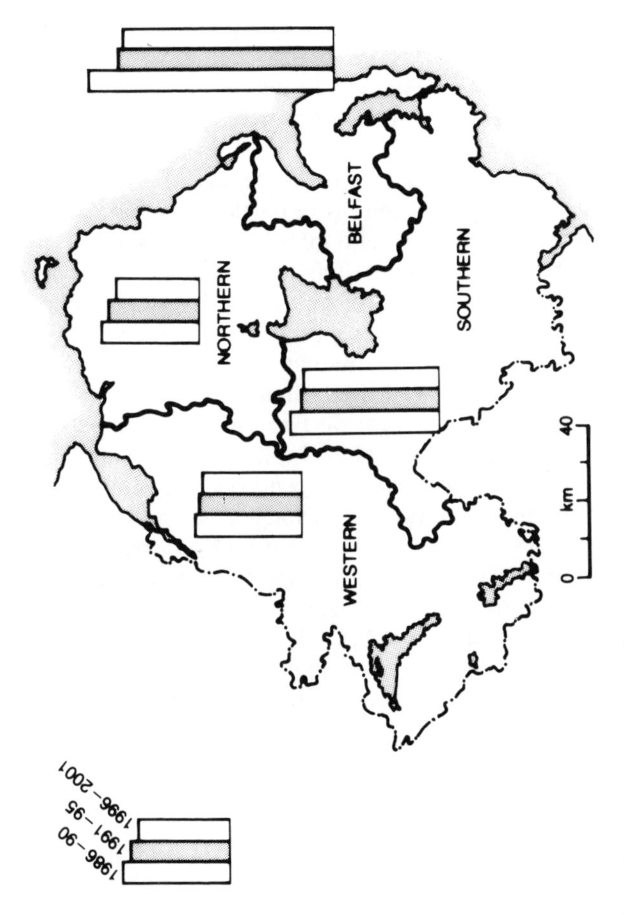

Figure 13 *Potential labour market entrants by area group, 1986–2000 (see also Table 4.1)*

Source: Own calculations based on *Northern Ireland Annual Report 1981* (General Register Office, 1984)

If one examines these changes over time at the district level, the disparities in prospects are even greater. Whilst Belfast shows a considerable reduction in the size of the entrant group, western districts like Omagh, Fermanagh and Cookstown would not experience as much of a reduction, unless a much larger proportion of young people now at school in these areas found it possible to migrate to another part of the Province, or overseas. Much will clearly depend on what will be the qualifications of local school-leavers, and their prospects of mobility, as well as on the size of the potential labour force leaver group.

In Tables 4.3 and 4.4 we examine, by district and by area group, the size and economic status composition of the age group 55–64 in 1981, that is those whom we can expect to stop working in the present decade, if they have not already done so. At this point we only look at the males, because female labour force participation rates in that age group are too small to make valid inferences possible. In Northern Ireland as a whole, 81.1 percent of males in that age group were still economically active, 68.4 percent being in work (either employed or self-employed) and 12.7 percent registered unemployed. This means that only just over two-thirds of that age group were occupying workplaces which would be vacated when they retired.

Looking at the local structure we find, as we would expect, that economic activity rates were highest in the Belfast group (84.4 percent), which also had the lowest percentages of unemployed, self-employed, and (prematurely) retired and permanently sick workers. From Belfast there is a steady gradient through our four area groups, with only 59 percent of all males employed (or self-employed) in area group IV (Western), and the highest percentage there of early retirers and permanently sick. At 24.5 percent the self-employed in that group form, for the UK, a very high proportion indeed: they are typically farmers, shopkeepers and artisans. We shall examine their prospects later.

Looking at individual districts we have, predictably, Castlereagh and North Down heading the list with respectively 86.2 percent and 84.3 percent economically active and only 6.7 percent and 6.8 percent unemployed. Other areas with high activity rates show higher unemployment rates: Newtownabbey, Larne, Lisburn and Carrickfergus have much above average unemployment rates. Belfast itself has high activity rates and average unemployment.

The lowest activity rates occurred in districts in the South and West, and are always coupled with high unemployment; Omagh shows the lowest activity rate (69.8 percent) and Strabane the highest unemployment in the older age group (20.3 percent). The highest figures for early retirers occurred in Cookstown and Newry and

Table 4.3 *Economically active and inactive males aged 55–64 years, by employment status and by district, 1981*

District	Economically active		Employed (incl. self-employed)		Self-employed		Unemployed		Economically inactive		Retired		Permanently sick		Overall male unemployment
	no.	%[1]	no.	%	no.	%	no.	%	no.	%	no.	%	no.	%	%
Antrim	1,216	79.4	1,051	68.6	232	15.2	165	10.8	315	20.6	87	5.6	228	14.9	15.4
Ards	2,215	83.7	1,936	73.1	402	15.2	279	10.5	432	16.3	162	6.1	269	10.2	11.3
Armagh	1,619	76.8	1,348	63.9	514	24.4	271	12.9	488	23.2	164	7.8	324	15.4	18.3
Ballymena	1,975	83.2	1,755	74.0	457	19.3	220	9.3	399	16.8	142	6.0	255	10.8	13.3
Ballymoney	816	77.6	657	62.5	201	23.0	159	15.1	235	22.4	58	5.5	177	16.9	21.1
Banbridge	1,125	80.9	965	69.4	306	22.0	160	11.5	265	19.1	88	6.3	177	12.8	14.7
Belfast	12,836	83.2	10,877	70.5	794	5.1	1,959	12.7	2,599	16.8	959	6.2	1,621	10.6	21.3
Carrickfergus	1,070	86.2	833	67.1	80	6.4	237	19.1	172	13.8	59	4.7	113	9.1	16.9
Castlereagh	3,018	86.2	2,183	79.4	237	6.8	235	6.7	485	13.8	205	5.8	279	8.0	9.1
Coleraine	1,530	79.6	1,320	68.7	354	18.4	210	10.9	391	20.4	152	7.9	238	12.5	19.7
Cookstown	837	73.8	640	56.4	274	24.2	197	17.4	297	26.2	109	9.6	188	16.6	28.9
Craigavon	2,278	82.3	1,952	70.5	331	12.0	326	11.8	491	17.7	144	5.4	341	12.3	19.0
Down	1,808	80.2	1,557	69.0	485	21.5	251	11.1	447	19.8	168	7.4	276	12.4	15.6
Dungannon	1,447	76.9	1,119	59.5	468	24.9	328	17.4	434	23.1	160	8.5	272	14.6	25.9
Fermanagh	1,933	77.2	1,578	63.0	835	33.3	355	14.2	572	22.8	219	8.7	353	14.1	21.5
Larne	1,206	85.2	1,038	73.4	218	15.4	168	11.9	209	14.8	89	6.3	119	8.5	16.2
Limavady	764	80.3	586	61.6	219	23.0	178	18.7	187	19.7	64	6.7	123	13.0	26.8
Lisburn	2,827	85.7	2,507	76.0	436	13.2	320	9.7	472	14.3	171	5.2	300	9.1	11.9
Londonderry	2,216	76.3	1,687	58.1	267	9.2	529	18.2	689	23.7	228	7.8	460	15.9	28.9
Magherafelt	1,011	74.2	776	57.0	332	24.3	235	17.3	351	25.8	112	8.2	239	17.6	26.1
Moyle	520	77.5	405	60.4	192	28.6	115	17.1	151	22.5	51	7.6	100	14.9	27.4
Newry and Mourne	2,246	75.1	1,691	56.5	569	19.0	555	18.6	745	24.9	281	9.4	462	15.5	30.9
Newtownabbey	2,738	86.9	2,426	77.0	233	7.4	312	9.9	411	13.0	157	5.0	254	8.0	12.5
North Down	2,456	84.3	2,258	77.5	294	10.1	198	6.8	457	15.7	229	7.9	228	7.8	7.8
Omagh	1,256	69.8	1,030	57.2	477	26.5	226	12.6	544	30.2	140	7.8	404	22.4	22.1
Strabane	1,081	74.9	788	54.6	338	23.4	293	20.3	362	25.1	96	6.7	266	18.4	32.4

[1] Percentage of all males 55–64 years in each district.

Source: Own calculations from handwritten tables supplied by Northern Ireland Census Office

Table 4.4 *Economically active and inactive males aged 55–64 years, by employment status and by area group, 1981 (percentages)*

Area group[1]	Econ. active	Employed	Self-employed	Unem-ployed	Econ. inactive	Retired	Perm. sick
I Belfast	84.4	73.4	7.7	11.0	15.6	6.0	9.6
II Northern	80.1	67.8	19.6	12.3	19.9	6.7	13.2
III Southern	78.2	63.8	20.3	14.4	21.8	7.7	14.1
IV Western	75.5	59.0	24.5	16.5	24.5	7.8	16.8
Northern (%)	81.1	68.4	14.4	12.7	18.9	6.8	12.1
Ireland (no.)	54,044	45,563	9,586	8,481	12,565	4,499	8,066

[1] For definition of area groups see Appendix A.
[2] % = percentage of all males 55–64 years in each area group.

Source: Own calculations based on information supplied by Northern Ireland Census Office

Mourne, and the highest proportions of permanently sick were to be found in Omagh (22.4 percent) followed by Strabane.

Self-employment varied from the very high proportions found in the mainly agricultural districts of Fermanagh (33.3 percent), Moyle and Omagh (more than 25 percent) down to the low figures of Belfast and its suburbs and Derry (all at 10 percent or less).

Although we should not make too much of local variations, the overall picture is clear. Apart from rural areas where real incomes may be sustained in the agricultural sector (but not universally so), the economic position of the pre-retirement group, in 1981, was not encouraging. Early retirement can mean a good pension, or savings, or a good living in the informal economy; but, when it is coupled with a high proportion of those registered as permanently sick, one must doubt this. In the last column of Table 4.3 we have recorded overall local male unemployment rates, and it is quite clear that low economic participation rates in the older age groups are strongly associated with high overall unemployment; this is hardly a feature which would make possible a comparison with those areas of Great Britain where there are also high early retirement rates (for example, in the South West region) but the picture is otherwise still very favourable.

Central Equation

In Tables 4.5 and 4.6 we give the results of what we regard as the central demographic equation of this report: the proportion of entrants to leavers in the labour market (see also Figure 14(a)–(c)). The detailed information on which these tables are based has already been presented in earlier tables.

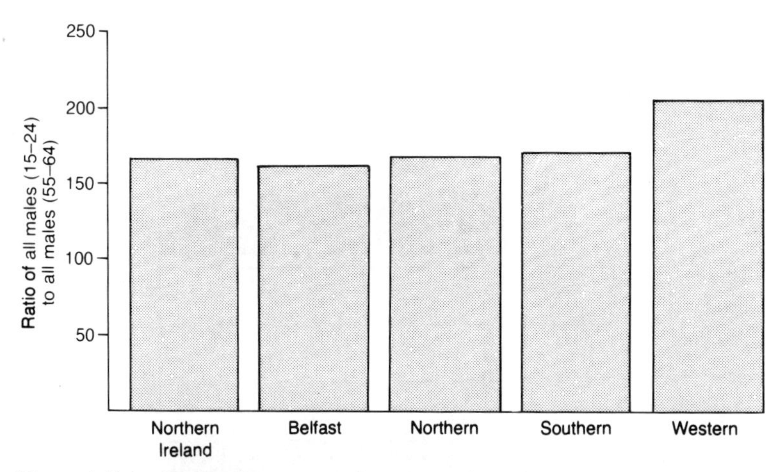

Figure 14(a) *Ratio of entrants to leavers, males only, 15–24 to 55–64, by area groups, 1971*

Source: Own calculations

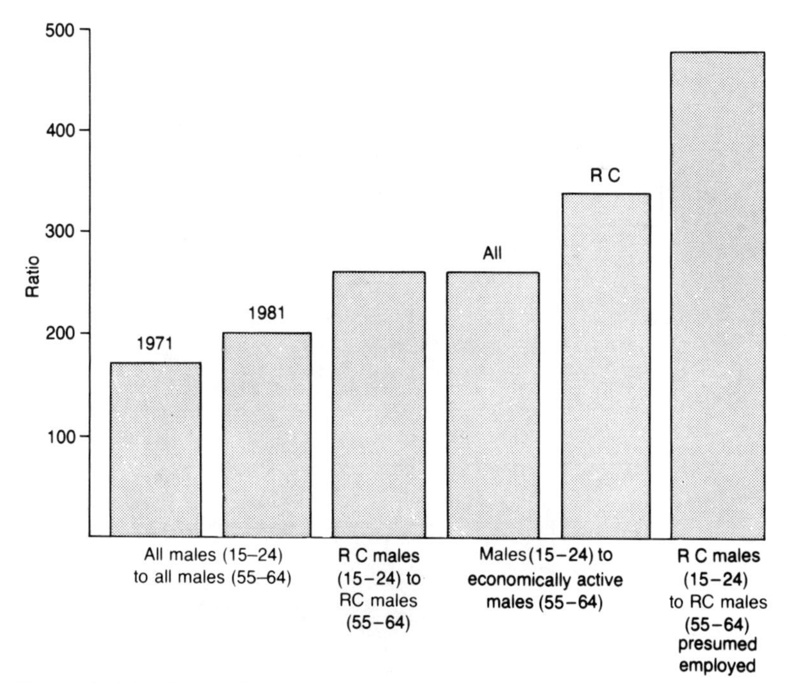

Figure 14(b) *Ratio of entrants to leavers, males only, by religion and economic activity status, Northern Ireland, 1981*

Source: Own calculations from Northern Ireland Census 1981

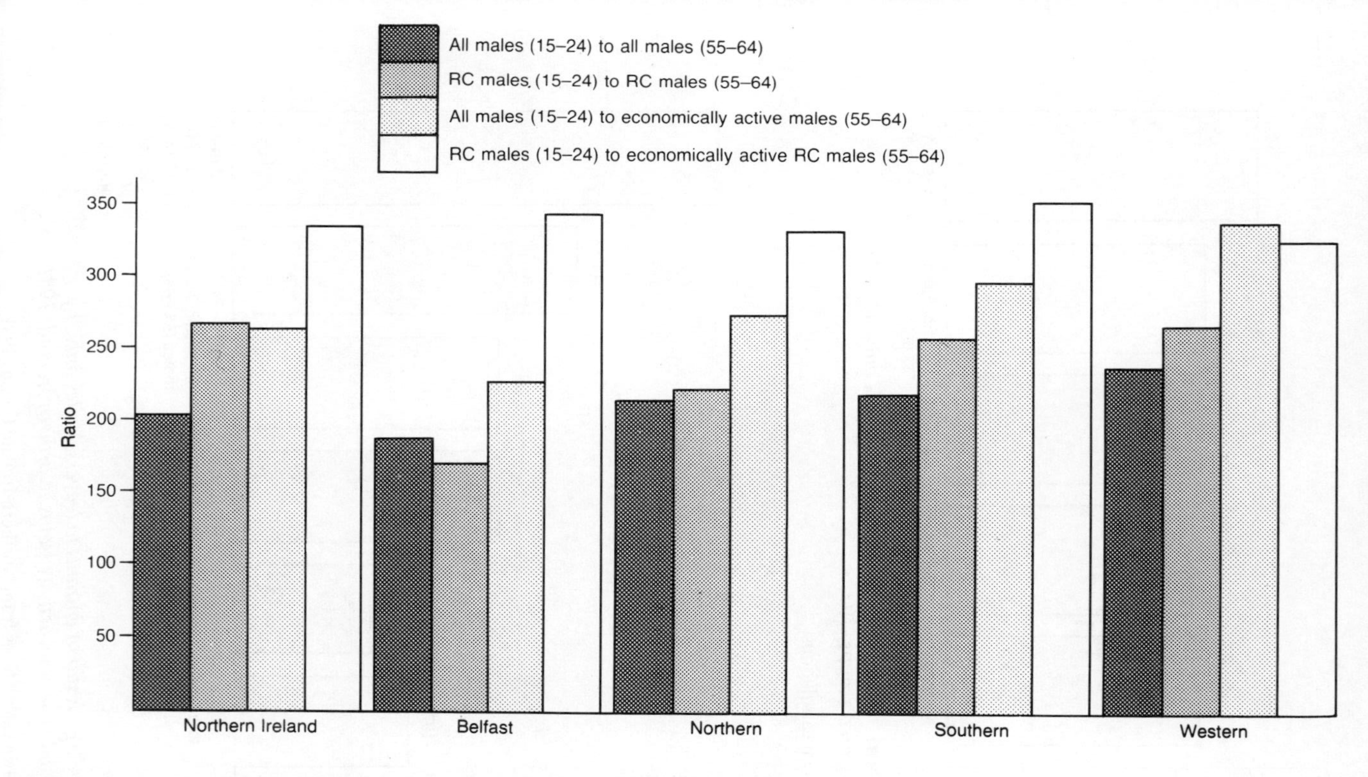

Figure 14(c) *Ratio of entrants to leavers, males only, by religion and economic activity status, Northern Ireland and area groups, 1981*

Source: Own calculations

The essence of this equation is that we must compare, for each period, the probable entrants to the labour force with those who are likely to retire from it. In a stationary economy and with a stationary population (that is, one which does not change in total size or age composition over time), there will always be just one entrant available to take up a workplace vacated by someone retiring from the labour market. Such a one-to-one ratio is in fact extremely unlikely to occur in practice. It is even more unlikely that there would be a correct match between entrants and leavers as regards their qualifications or preferences. The smaller the labour market, the more unlikely is it that this ratio of one to one can be achieved.

Nevertheless, we may use it as a benchmark. Any economy which is not subject to an overall reduction of labour demand will suffer a disadvantage if there is not one entrant for every leaver – a situation which was sometimes described in the 1930s, in Britain, when the birth rate had fallen very sharply and the country was experiencing a return to full employment. Conversely if fertility has been relatively high, or if the leaver age groups have been eroded by other factors (e.g. emigration), then there will be more than one entrant for every retirer. The arithmetic is somewhat theoretical, but as a measure of comparison between labour markets, and in any one area over time, it is a useful device for assessing at least one of the causes of unemployment, and for forecasting likely future levels of unemployment.

The method does not take into account various other reasons why young entrants may find it impossible to obtain jobs: for example, sharply rising productivity, change to capital-intensive production, lack of qualifications and so on. We shall deal with these matters later.

For the basis of our calculations, we have here compared ten-year age groups in 1981: those aged 15–24 years, and those aged 55–64 years. The first group will have been actively seeking work in 1981, or will have become entrants during the 1980s if they were still in full-time education or training in 1981. All of them will be potential workers by 1991. The older age group, in 1981, will have included some who had already retired from the labour market, and practically all of them will have withdrawn by 1991. The ratios therefore relate to a decade over which a continuous process takes place, and cannot be used to 'explain' unemployment in a single year. They can serve, however, to illustrate time trends and to make interareal comparisons.

In this section we also introduce, for the first time, religious affiliation, to illustrate another area of comparison. We also calculated the same basic ratios for 1971, and (except for the religious affiliation) for the rest of the UK for 1981.

Table 4.5 *Ratios of male entrants and leavers groups, by district, 1971 and 1981*

District	Ratio of all males 15–24 to all males 55–64		Percentage increase/ decrease in ratio	Ratio of RC males 15–24 to RC males 55–64, 1981 Census
	1971 Census	1981 Census		
	1	2	3	4
Antrim	193	294	52.3	354
Ards	148	160	8.1	213
Armagh	170	199	17.1	234
Ballymena	153	187	22.2	282
Ballymoney	162	200	23.5	286
Banbridge	139	190	26.8	250
Belfast	162	182	12.3	267
Carrickfergus	200	198	−1.0	297
Castlereagh	160	149	−6.8	222
Coleraine	196	222	13.3	267
Cookstown	169	207	22.5	243
Craigavon	181	236	30.4	356
Down	165	204	23.6	258
Dungannon	166	204	22.9	231
Fermanagh	143	190	32.8	186
Larne	162	190	17.3	264
Limavady	210	291	38.6	295
Lisburn	158	237	50.0	328
Londonderry	211	279	32.0	354
Magherafelt	171	207	21.1	237
Moyle	158	184	16.5	182
Newry and Mourne	180	236	31.1	243
Newtownabbey	203	193	−4.9	309
North Down	145	172	18.6	284
Omagh	151	214	41.7	209
Strabane	171	224	31.0	251
Northern Ireland	166	204	22.9	264

[1] Column 7 denotes economically active men less those returned as unemployed or chronically sick.

We note the deterioration in the previous decade: in 1971, there were 166 men in the entrant group for every 100 leavers; in 1981 there were 204, a deterioration of 22.9 percent. To that extent, the worsening of the unemployment situation, especially among the young people, can be attributed to 'purely' demographic forces. However, as we showed in Chapter 2, this is not in fact a matter of large numbers of children only; it is also connected with the 'hollowed-out' generations in the older age groups, thanks to previous emigration.

The position in the labour market is in fact even worse than that,

Table 4.5 *continued*

District	Ratio of males 15–24 to economically active males 55–64, 1981 Census		Ratio of RC males 15–24 to RC males 55–64 presumed employed[1]
	All men	RC men	
	5	6	7
Antrim	377	460	608
Ards	190	231	293
Armagh	294	339	476
Ballymena	226	298	383
Ballymoney	256	342	489
Banbridge	234	272	353
Belfast	236	331	480
Carrickfergus	230	256	322
Castlereagh	174	233	255
Coleraine	282	341	471
Cookstown	309	379	669
Craigavon	295	315	453
Down	279	362	451
Dungannon	290	343	514
Fermanagh	248	275	393
Larne	225	270	357
Limavady	372	435	687
Lisburn	283	397	510
Londonderry	414	461	503
Magherafelt	304	355	521
Moyle	238	298	431
Newry and Mourne	280	315	488
Newtownabbey	223	284	348
North Down	205	251	281
Omagh	343	386	530
Strabane	314	359	588
Northern Ireland	263	336	481

Sources: *Northern Ireland Annual Report 1981* (General Register Office, 1984) and Northern Ireland Census 1981, *Religion Report*. Own calculations based on 1981 County Reports, Tables 7 and 12

because really we should add the women in the lower age groups to the entrants. Increasingly, compared with the past, women are competing with men in many occupations and, among those retiring, there will be a good many men who will be replaced by women: shop assistants, local and national government officers, clerical workers and school teachers. However, we cannot guess at the proportion. Similarly, a high proportion of the women entrants will replace other women who retire to raise a family, or for other reasons, and there is

Table 4.6 *Ratios of male entrants and leavers groups, by area group, 1971 and 1981*

	Ratio of all males 15–24 to all males 55–64		Percentage increase/ decrease in ratio	Ratio of RC males 15–24 to RC males 55–64, 1981 Census	Ratio of males 15–24 to economically active males 55–64, 1981 Census		Ratio of RC males 15–24 to RC males 55–64 presumed employed[2]
Area group[1]	1971 Census 1	1981 Census 2	3	4	All men 5	RC men 6	7
I Belfast	163	185	13.5	169	226	343	447
II Northern	169	214	26.6	220	272	330	472
III Southern	170	215	26.4	259	296	352	477
IV Western	205	238	16.1	264	338	323	530
Northern Ireland	166	204	22.9	264	263	336	481

[1] For definition of area groups see Appendix A.
[2] Column 7 denotes economically active men less those returned as unemployed or chronically sick.

Sources: *Northern Ireland Annual Report 1981* (General Register Office, 1984) and Northern Ireland Census 1981, *Religion Report*. Own calculations based on Northern Ireland *Annual Abstract of Statistics* no. 3, Table 8

no way of gauging this effect. So we have to be content with the statement that the proportion of males understates the situation, though we can still make comparisons over time, between areas within Northern Ireland, and with Great Britain.

We now further refine the calculation. First we look at Roman Catholic males only; the situation is that there are 264 entrants for every 100 leavers (column 4 in Tables 4.5 and 4.6). If we now take out of the older age group those whom we know to be economically inactive (as shown in the previous section) we obtain proportions of 263 for all men and 336 for Roman Catholics (columns 5 and 6). Finally, if we further refine the figures for the Catholics by removing those who are economically active, but unemployed, the proportion is 481 to 100, or nearly 5:1. (These figures are based on the economic activity tables of the Census religion data, and exclude the 20 percent 'not stateds'. However, as we demonstrated at the start of this report, a majority of non-respondents were Roman Catholic, and a higher proportion of them unemployed, early retired, sick, etc.; so all we would get from dividing up the 'not stated' population is still more extreme disproportions.)

It will be seen by reference to Table 4.7 that in the rest of the UK we also have a deterioration between 1971 and 1981, from 130 to 150, but this is still a great deal more favourable than Northern Ireland (see also Figure 15). In column 5 of Table 4.7 we show what happens if we perform the same calculation as the one which gave us column 7 in the previous table, that is take out the economically inactive and the unemployed, and we get a figure of 200. If we repeat the operation for a few selected areas of Great Britain, we find the South East region better off (if not overall, at least using the measure of the last column), and some of the black spots of the British economy also showing up rather badly, with figures in the 200s being general. If we take Knowsley in Merseyside, with its young population in public sector housing, we obtain a figure of 343 – probably the worst in Great Britain, but much better than that for Northern Irish Catholics.

If we now compare the area groups within Northern Ireland (Table 4.6) we have our usual gradient, which applies in all categories except column 6, where the Northern and Southern groups for once are worse off, apparently, than the Western group. But the overall picture remains the same. The Western group has proportions, for all denominations, rivalled only by Liverpool in Great Britain.

These overall effects are clearly still more marked at the most local level, and it is here that we have eleven districts where the proportions for Catholic men are about 5:1 or worse (Table 4.5). It should be noted that this applies not only in the Western area group, but also in some largely Protestant districts: Antrim, Ballymoney, Coleraine,

Table 4.7 *Comparative data on ratios of entrants to leavers for the UK, 1971–81*

	Ratio of all males 15–24 to all males 55–64				
	1971 Census	1981 Census	Percentage change in ratio 1971–81	Ratio of all males 15–24 to economically active males 55–64, 1981	Ratio of males 15–24 to economically active males 55–64 presumed employed, 1981[1]
	1	2	3	4	5
Great Britain	130	150	15.4	179	200
England and Wales	129	148	14.7	177	197
South East region	—[2]	153	—	174	188
Birmingham district	—	153	—	179	214
London: Lambeth	—	179	—	203	223
London: Brent	—	188	—	209	227
London: Hackney	—	179	—	204	228
Liverpool district	—	172	—	209	251
Knowsley district	—	227	—	281	343

[1] Column 5 denotes economically active men less those returned as unemployed or chronically sick.
[2] Indicates figures unavailable.

Source: Own calculations based on 1981 County Reports, Tables 7 and 12

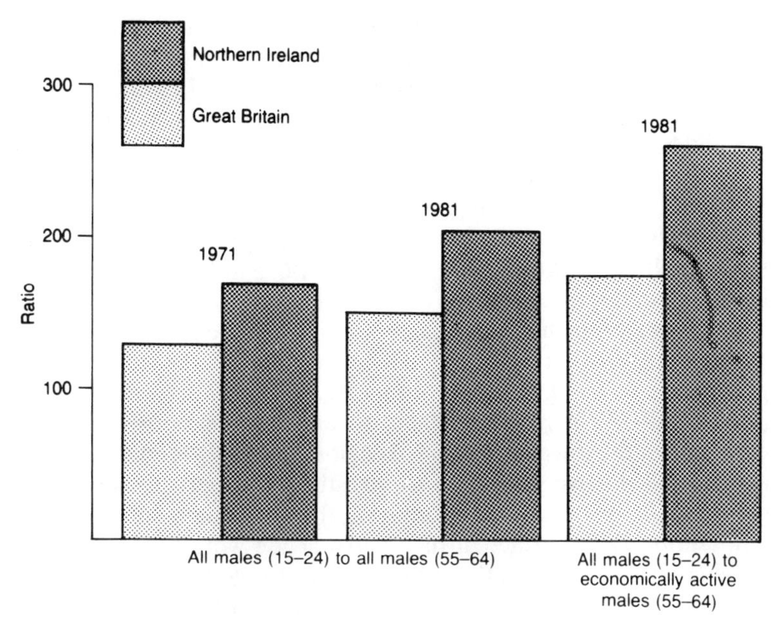

Figure 15 *Comparison for Northern Ireland and Great Britain between size of entrant and leaver groups in the labour market on various assumptions, 1971 and 1981 (see also Table 4.7)*

Sources: Own calculations based on 1981 county reports, Tables 7 and 12; Northern Ireland *Annual Abstract of Statistics* no. 2, Table 8

Lisburn, and of course Belfast itself. Just how local are these labour markets we do not know. We have not been able to test the relationship for travel-to-work areas over and above a crude comparison of these ratios with local religion-specific unemployment rates. The agreement is complete, so we can say that these district figures represent the true chances of employment for labour force entrants at least in the medium run. No doubt if we were able to make comparisons at an even finer grain (for example, West Belfast, or the NIHE estates in Lisburn) we might arrive at proportions of up to 10:1, but it is really not profitable to pursue the matter to a *reductio ad absurdum*.

It should be noted that even the favourably situated areas, for example in the Belfast outer ring, do not show brilliant prospects, but at least they are no worse than the old British industrial regions. Ards, Castlereagh and North Down clearly head the list, and still do so even when early retirement is taken into account (in column 5), and the position for Roman Catholics in these areas is again not too serious by comparison with Cookstown or Limavady. As we shall show, this is a function of occupational and industrial structure, and cannot be put down to the small proportion of Roman Catholics in these districts. We have shown elsewhere that in the Province, where there are relatively small Catholic minorities, religion-specific ratios can still be very bad indeed.

To sum up the implications of these calculations: other things being equal, we would expect only a small minority of boys, and Roman Catholic boys in particular, to obtain work within a few years of leaving full-time education, wherever they live in Northern Ireland, in the next decade. Exactly who will get any available jobs will depend on a large number of factors which may be interrelated: the nature of the work, the educational qualifications of the entrants, the proportion of public sector workplaces, and local access problems (for example, from housing estates to industrial areas possibly at the other end of the town or district). Religion, however, is bound to play a part, and this we shall have to investigate further. 'Other things being equal' means, first and foremost, that the total number of jobs is not seriously diminished further, and that new workplaces continue to be created to balance those that are lost for good. The unknown variables will also include the proportion of girls obtaining posts formerly held by men, instead of 'women's work' only.

Subject to all these cautions, we can read the figures as meaning that overall, in the 1980s, 50 percent of all school-leavers will not obtain work, and that for some sections of the population, and in some places, that proportion will rise to 80 percent and more. As we shall see when we analyse unemployment figures, these calculations are not unrealistic.

Such calculations have also been made for areas of high unemployment in Great Britain. They have been the main reason for the large number of schemes now in operation, or proposed for the future, designed to keep young people under 20 out of the labour market altogether by more full-time education and training, and the various Manpower Services Commission schemes. If the demographic structure were fixed, such measures would merely postpone the onset of unemployment, apart from the slight reduction in the total number of people who offer themselves on the labour market at any one time. As we have demonstrated, however, the demographic situation is not static, and thus the proportions given in our tables represent the position in the 1980s only, with some improvement likely to occur from the mid 1980s onwards.

It would be tempting to project these ratios for the next fifteen years. As we have seen, we have a fair idea of the number of entrants, subject only to migration. What about the leavers? As Table 1.5 showed, on current projections there will be some increase in the size of that group, mostly towards the end of the century, reflecting the larger size of the male age groups in the younger adult ages shown in the 1981 Census, especially those in their late 30s who will all be approaching retirement by the end of the century. Given the probable decline in the entrant groups this should, once again all other things being equal, refer to a better proportion between entrants and leavers.

However, this assumes that among these larger older groups, fifteen years hence, there will be as high a proportion economically active and working as there is now. On this we cannot make any prognostications. There is a rising tide of pressure for earlier retirement; quite possibly some new pensions regulations are in the offing. Unemployment rates have been rising even for the middle group of adults; why should we assume that they will have found work by then? Is there any indication that those who describe themselves as permanently sick will decrease over time? Given these uncertainties, we have not carried out a projection of entrants/ leavers ratios.

If we tried to break down the future entrants, at least by religion, we could be on slightly firmer ground: we do have a fairly good idea to what denominations the entrants in the late 1980s and the 1990s will belong, because we have already calculated the denomination structure of the under 15s in 1981 (Eversley and Herr, 1985). Because the lower fertility rates of the Roman Catholic population will not work their way through into the labour market in significant proportions until the next century, we have to take the percentages calculated from the 1981 school count as reflecting the true proportions of future entrants (assuming no denominational bias in emigration): 46.2 percent of the 15 year olds in 1981 (nearly all on the labour market by the time this report appears), 45.4 percent of the 10 to 14 year olds in 1981 (who will all enter the market in the 1980s and early 1990s), and 46.3 percent of 5 to 9 year olds in 1981 (all labour market entrants before the end of the century). But only 27 percent of the over 50s were Catholics in 1981 (Eversley and Herr, 1985). Thus one has to guard against too much optimism as regards the future amelioration of the demographic situation.

Again local projections are even more hazardous than national ones, but we already know that two-thirds of all school children were Catholics in our Western area group in 1981, and 54 percent of the over 50s. This is perhaps a marginally better outlook compared with the Province as a whole, if we forget about the lower proportion of those actually working in the older age groups.

For all these reasons, we think it is best to hold on to the relatively firm ground of the statistics presented in Tables 4.1 and 4.2 to show the prospects in different areas, and to the warning that things may get still worse.

Total Labour Market Changes 1971–81, and Prospects

In the preceding sections we have tried to estimate total population changes for the decade 1971–81 (by a simplified analysis of the relationship between the size of the entrant and the leaver groups), and to extract economic participation and unemployment data. We have also considered the overall probability of young entrants to the labour market finding work in the current decade. However, we have taken into account, in these calculations, only inactivity and unemployment in the older age groups.

It is now convenient to summarize the total labour market changes, at least for males; to reassess the starting position, as it were, from which the probable future movement of employment might be estimated.

In the summary in Table 4.8 we have taken a combination of corrected Registrar General's midyear population figures (to arrive at the size of the total male age groups concerned), Census statistics (to obtain the proportion of economically inactive males), Department of Economic Development statistics (to check on the number of registered unemployed), and OPCS statistics (for the age and sex structure of the out-migrants). It is realized that these different sources are not totally compatible: nevertheless, as a method of assessing the total labour market failure in the decade, the figures serve well enough. The lines in the table are self-explanatory. Reduction of employees in employment and self-employed measures the 'economic labour gap', that is the net loss of sources of gainful employment (lines 1 + 2 = line 7). Lines 3 + 5 (= line 8) show to what extent this loss of workplaces is reflected in registered unemployment and rise in inactivity (the latter consisting, apart from a student element, mainly of prematurely retired persons and permanently sick). If we then add those adult males whom we believe to have migrated out of Northern Ireland (lines 8 + 9) we obtain the total labour market failure in the decade; that is, the number of jobs that would have had to be created to leave unemployment and economic participation rates in 1981 no worse than they were in 1971, without further out-migration.

We can update these figures in part at any rate to 1984, always remembering that non-Census sources are not always fully compatible with Census estimates. We have added columns to show changes after 1981 where comparable statistics are available, or informed guesses can be made.

Table 4.8 *Summary of labour market changes, Northern Ireland, 1971–81 and 1981–84, males only*[1]

	1971	1981	Change 1971–81	1984	Change 1981–84	Change 1971–84
1 Employees in employment	290,700	268,650	−22,000	245,750	−23,000	−45,000
2 Self-employed	74,000	69,000	−5,000	69,000	nil	−5,000
3 Unemployed	30,400	70,000	+39,600	90,000	+20,000	+59,600
4 Working population (1 + 2 + 3)	395,250	407,650	+12,400	404,750	−3,000	+9,400
5 Economically inactive aged 16–64	58,700	72,800	+14,100	—	—	—
6 Population aged 16–64 (4 + 5)	452,000	470,340	+18,000	495,000	+25,000	43,000
7 Loss of employment and reduction of self-employment (economic labour gap) (1 + 2)			−40,650		−23,000	−63,650
8 Increase in registered unemployment and in men of working age economically inactive			59,000		20,000[2]	79,000
9 Estimated out-migration (net) of working age males			32,600		7,500	40,000
10 Total labour market failure (8 + 9)			91,340		27,500	119,000

[1] Figures have been rounded where based on estimates rather than counts.
[2] Unemployed only.

Sources: Own calculations based on Northern Ireland Census 1971 and 1981 economic activity tables, Registrar General's estimates, Northern Ireland *Annual Abstract of Statistics* no. 3, and OPCS migration estimates 1981–85

Male employees in employment decreased by a further 23,000 from 1981 to 1984. Self-employed males remained more or less static. The unemployed, on the basis of claimants counted, increased by a further 20,000 (see Chapter 5). The total working population continued to increase, but we cannot say by how much: each year about 15,000 boys reached working age in the early 1980s, but we do not know how many of these remained in education, training or on special employment schemes outside the normal labour market. That was true in the previous decade also, and so perhaps we should not make any special adjustment for these factors.

At the same time, another 6,500 males would have reached official pensionable age annually, so that the cumulative increase for the years 1981–84, in the population of working age, would be about 25,000.

We do not know, in fact, how many males in this age group (15–64) were economically inactive (either because they were still in education or training, or because of early retirement), so that we have to base the totals in line 8 only on the known unemployed population. On these calculations, the increase in unemployment is about the same as the increase in the potential workers, without allowing for the out-migrants in line 9.

To summarize the further changes in the last few years, it is possible to guess that the deterioration in unemployment is in large measure accounted for by a change in the size of the potential working population. The increase in the number of unemployed would have been greater had it not been for continued net out-migration, and, one must assume, it would also have been greater had it not been for increases in educational participation rates in the young population, and the special training schemes in operation. One may also safely assume that more of the older workers retired early from the workforce, and are not recorded as benefit-claiming unemployed.

As regards those creamed off from the potential young unemployed by various government schemes, these have been described in detail in successive notices issued by the Department of Economic Development (for copies of specimen notices see the Additional Tables). The effect of these schemes can be gauged by the fact that unemployment of the under 20s (males only) rose from 7,350 in January 1980 to 12,512 in January 1981, having reached 14,193 (including the new school-leavers) in July 1980. Since then the number of young unemployed (registered claimants) has fallen absolutely, and certainly relative to the size of the age group. In July 1981, just after the Census, 14,324 of the 16 to 19 year olds were unemployed, that is 23 percent of those estimated to live in the Province in mid 1981. By July 1984 that figure had dropped to 12,440, or 19.7 percent of the relevant age group (that is those who were aged 13 to 16 years in 1981). That suggests that (allowing for the general reduction in jobs) some 2,500 youngsters were removed from the labour market, and this accords quite well with the official account. This rough calculation shows how hard it will be to prognosticate registered unemployment levels in general, let alone by sex, age, district and religion.

All these calculations are subject to a margin of error. Not only are there discrepancies between the Census employment figures and those recorded by the Department of Economic Development, but there are further difficulties which become apparent if one takes the Labour Force Survey as the basis for the analysis. Recent Labour Force Survey results suggest that the 'true unemployed' (that is those actually wanting to work) are not so very different from the 'claimant-based' unemployed – probably because registered unemployed not really wanting to work are cancelled out by people who would like to work but are not registered as unemployed.

These difficulties do not in fact alter the general picture presented in Table 4.8, or the time trends, except in the sense that the cumulative 'discouraged worker' effect, plus the increase in various activities which keep the 16 to 19 age group out of the labour market, reinforce the widely held view that the official figures somewhat understate the true extent of the problem. This, however, is not the main question addressed in this investigation.

It would therefore be idle to speculate on what would happen in the next five years, let alone how such changes would be distributed as between regions and districts. Critics have expressed doubts about the Department of Economic Development's 'employee' base which has been irregularly revised from year to year. The fact is, however (as work on the size of the unemployment problem in Great Britain has also shown), that it is difficult to estimate accurately, on an annual basis, how many persons were actually working, plus those actively seeking work, to form the denominator for the ratio which gives us unemployment rates. Clearly, if people on temporary employment schemes are excluded, and also those no longer registering as unemployed (that is those over 60 under the new regulations which were introduced in 1982), it becomes quite possible for the 'potential working population' not to increase substantially despite the demographic increase, and for the unemployed in the younger age groups also to remain fairly stable. The latter possibility is in fact confirmed by the Department of Economic Development's annual estimate of the age structure of the unemployed, which shows increases in the young unemployed group much smaller than would have been expected by looking at the size of the age group in the midyear estimates.

These difficulties, however, cannot be further discussed in this report: they relate to administrative procedures and benefit legislation rather than the analysis of the labour market.

To sum up, although it is not possible to demonstrate a further increase in the rate of labour market failure for the period 1981–85, analogous to that which occurred in 1971–81, we note three undisputed facts:

1 The number of unemployed has continued to rise.
2 The number of those recorded as being in employment has continued to fall.
3 The number of potential labour market entrants is falling only slowly (the 16 year olds having reached their peak in 1981) and continues to exceed the number of potential leavers by a very substantial margin. Emigration does continue, and *may* still provide the equilibrating factor between fall in employment and rise in unemployment. However, we can make no predictions as to the future course of net emigration.

We therefore conclude this chapter by stating that we can see no prospect of the 1981 situation improving in the foreseeable future, and every likelihood of further, though probably smaller, increases in unemployment occurring.

5

Employment Status of the Northern Irish Population in 1981: Area and Religion

Overview

We must now turn to the differentials in employment and occupational status observed in Northern Ireland in 1981, to show the differentials which occur by area and by religious affiliation. We have to rely, for this and the following chapter, almost entirely on the results of the 1981 Census. We are well aware of the difficulties presented by the 'not stated' category. As we indicated in the Introduction, we are proceeding on the assumption that any resolution of the 'not stateds' into their proper denominational groups could only exacerbate the contrasts we shall present. In other words, the fact that those who did not answer that question tended to be more Catholic than Protestant (the 55:45 division implying a disproportionately larger addition to the Catholic respondents than to the Protestant ones), and the known propensity of the 'not stateds' to belong to lower-status groups, all point the same way: if we could allocate the remainder, the differentials would be increased. In some commentaries on the tables, we shall point out what would happen if we did make assumptions about the divisions of the 'not stateds', but this will, as a rule, be unnecessary.

In Tables 5.1 and 5.2 we show the general and area group structure of the population. In the Province as a whole, just over 10 percent of all adult men were self-employed, 9 percent had managerial or foreman status, and under 44 percent were other employees; 15 percent were unemployed. For women, neither self-employment nor higher employed status was at all significant: almost all those who worked were rank-and-file employees.

In Table 5.2 we look more closely at economic activity rates by religion and by area group. It is at once apparent that overall activity rates for Catholics (both men and women) are not significantly different from those for the whole population. (Since Roman Catholic inactivity rates are slightly higher, it follows that if 'all denominations' were broken down to show all stated religions other than Catholics, that is, mostly Protestants, the difference would be slightly greater, but the totals show that this does not amount to much.) It is

Table 5.1 *Employment status of Northern Ireland males and females, 1981*

Status	Males no.	Males %	Females no.	Females %
Self-employed with employees	15,242	2.97	2,981	0.54
Self-employed without employees	37,354	7.28	2,106	0.38
Managers	25,325	4.94	6,015	1.08
Foremen	20,294	3.96	11,030	1.99
Others	223,900	43.64	180,047	32.40
Total employees	269,519	52.53	197,092	35.47
Out of employment	76,269	14.87	29,196	5.25
Economically active	398,384	77.65	231,375	41.64
Total population 16+	513,043	100.00	513,043	100.00

Source: Own calculations based on Northern Ireland Census 1981, *Economic Activity Report*

interesting to note at this stage that a higher proportion of Catholics are returned as students, both males and females. Unfortunately, married women are not separately distinguished in the religion printouts of the economic activity tables, so we only have general figures for female activity rates, overall and part time.

Looking at area figures, we find the expected differences, but more especially in women's activity rates (see also Figure 16). The Belfast group has the highest women's participation rate, at 44.6 percent (38.8 percent of married women), and the Western area group has the lowest rates: 37.2 percent of all women (34.1 percent of married women).

Figures for Catholics reflect the area patterns: participation rates are about the same, for men and for women, for the denominations in all area groups, reflecting the general gradient from east to west (see Figure 17). There are small variations in the relative proportions of all women working, and married women working, which may or may not be reflected in the part-time economic activity rates of Catholic women: the differences are not large enough to be significant. Married women part-time workers are noticeably more numerous in the Belfast group, and lower in the Western group, but this, as we shall see, almost certainly reflects the industrial structure of the areas, rather than religion-specific differences in attitudes to married women working, or even fertility. From district figures we can see that, even in predominantly Catholic areas, married women work if there are the jobs available; this is what we would expect on the basis of British labour market studies, which show that the proportion of part-time women working is affected not by family size differentials but by a mixture of need and opportunity.

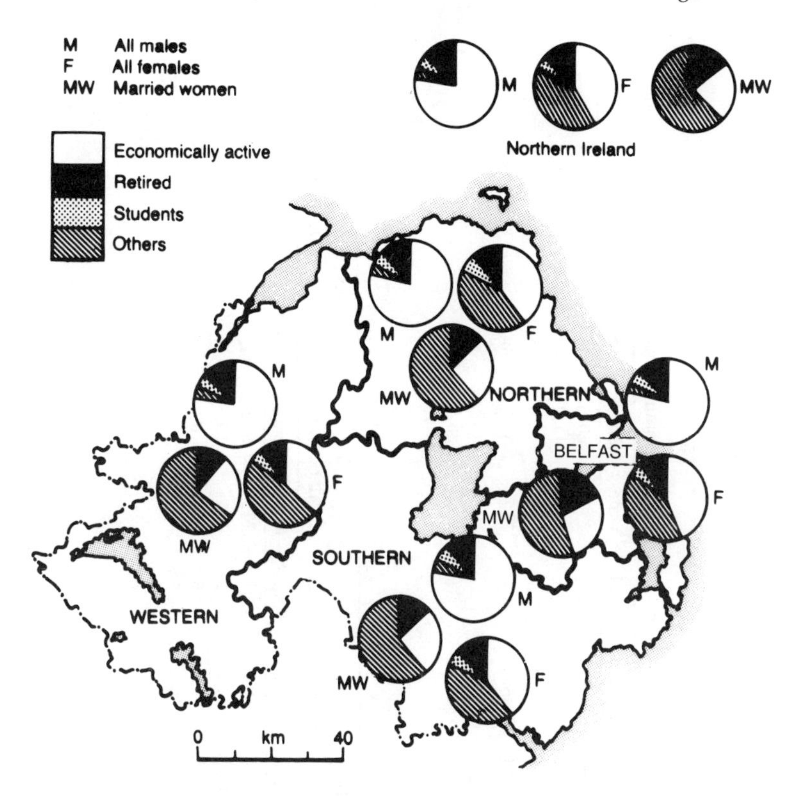

Figure 16 *Occupational structure, Northern Ireland and area groups, 1981 (see also Table 5.2)*

Sources: Own calculations based on Northern Ireland Census 1981, economic activity tables, Table 14b (unpublished computer printout) and Northern Ireland Census 1981, *Economic Activity Report*, Table 4

Once again, we find the student differential: 8.4 percent of Catholic males over 16 were students in the Belfast area group, a very high proportion, and 8 percent of all women in the Northern area. There was, in 1981, nothing in the figures to suggest that either Catholics, or women, were less likely to be pursuing their studies beyond minimum school-leaving age than non-Catholics, or men; or if there was, it was only by very small margins indeed.

How are we to interpret these tables? If anyone were to assert that Roman Catholics were less likely to want to work, or less likely to take advantage of educational opportunities, we could reply that none of this is true, at least on the basis of the figures given here. Could these proportions be affected if we were to analyse the 'not stateds'? It seems highly unlikely. We have looked at the proportions

Table 5.2 *Occupational structure of Northern Ireland, males and females, all denominations and Roman Catholics only, by area group, 1981 (percentages)*[1]

	Group I[2]						Group II					
	All denominations			Roman Catholic			All denominations			Roman Catholic		
	Male	Female	Total	Male	Female	Total	Male	Female	Total	Male	Female	Total
Economically inactive	21.7	55.4	39.6	21.9	54.8	39.6	22.0	60.6	41.0	23.7	61.5	43.2
Retired	12.6	11.9	12.2	9.2	11.6	10.5	12.5	11.2	11.8	11.1	10.7	10.9
Student	6.0	5.0	5.5	8.4	7.0	7.6	5.6	5.9	5.7	7.6	8.0	7.8
Economically active	78.3	44.6	60.0	78.1	45.3	60.4	78.0	39.4	58.1	76.3	38.5	56.8
Married women economically active[3]		46.0						38.9				
Married women economically active part-time		19.0						13.3				

	Group III						Group IV					
	All denominations			Roman Catholic			All denominations			Roman Catholic		
	Male	Female	Total	Male	Female	Total	Male	Female	Total	Male	Female	Total
Economically inactive	23.0	60.2	42.0	23.8	60.6	42.7	23.7	62.8	43.3	25.0	63.0	44.4
Retired	12.6	12.1	12.4	11.5	11.6	11.5	12.6	10.7	11.6	12.4	11.0	11.7
Student	6.1	6.3	6.2	7.2	7.5	7.3	6.0	6.8	6.4	6.7	7.5	7.1
Economically active	77.0	39.8	8.0	76.2	39.4	57.3	76.3	37.2	56.7	75.0	37.0	55.6
Married women economically active		38.8						34.1				
Married women economically active part-time		13.1						10.5				

Table 5.2 *continued*

		Northern Ireland					
		All denominations			Roman Catholic		
		Male	Female	Total	Male	Female	Total
Economically inactive	(no.)	114,659	324,292	438,951	31,755	87,157	118,912
Economically inactive	(%)	22.3	58.4	41.1	23.6	59.6	42.4
Retired	(no.)	64,614	64,733	129,347	14,881	16,539	31,420
Retired	(%)	12.6	11.6	12.1	11.1	11.3	11.2
Students	(no.)	30,549	31,577	62,126	10,028	10,812	20,840
Students	(%)	6.0	5.7	5.8	7.5	7.3	7.4
Economically active	(no.)	398,384	231,375	629,759	102,842	59,003	161,845
Economically active	(%)	77.7	41.6	58.9	76.4	40.4	57.6
Married women economically active	(no.)		120,101				
Married women economically active	(%)		37.8				
Married women economically active part-time	(no.)		49,411				
Married women economically active part-time	(%)		15.5				
Total population		513,043	555,667	1,068,710	134,597	146,160	280,757
Total married females			318,148				

[1] All percentages calculated as proportions of persons 16+, male or female.

[2] For definition of area groups see Appendix A.

[3] Married women calculated as proportions of all married women only. This category was not recorded by religion in the unpublished computer printout.

Sources: Own calculations based on Northern Ireland Census 1981 economic activity tables, Table 14b (unpublished computer printout), and *Economic Activity Report*, Table 4

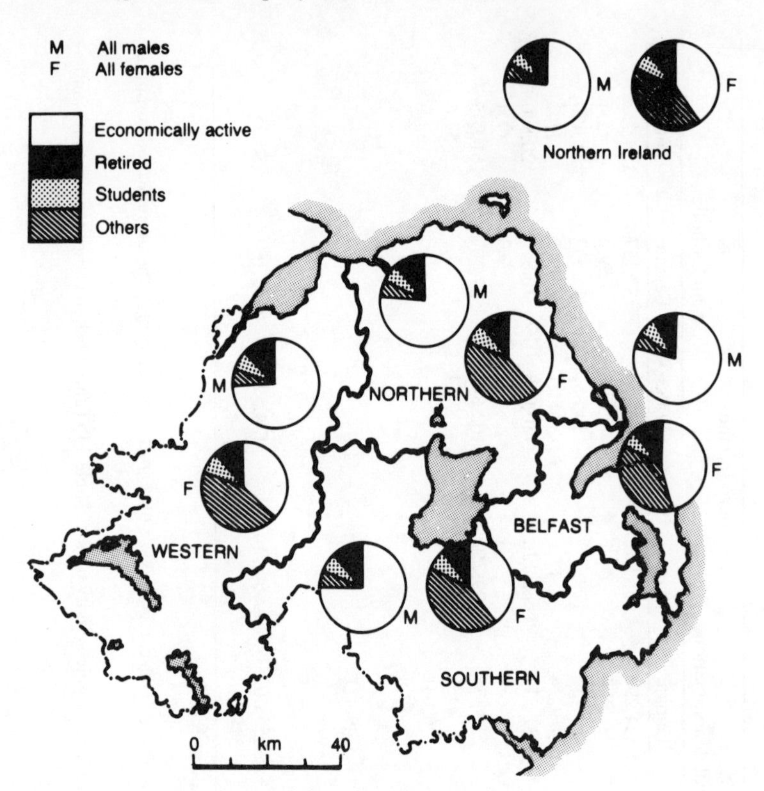

Figure 17 *Roman Catholic occupational structure of Northern Ireland, by area group, 1981 (see also Table 5.2)*

Source: Own calculations based on Northern Ireland Census 1981 and unpublished computer printout

of known Protestants and the 'not stateds' in the economic activity tables of the religion reports, and they do not differ significantly from the whole population, or Catholics. A slightly higher proportion of women were economically active among the 'not stateds' than among the whole population; the proportion of students was nearer the 'all denominations' figure than the Catholic one, but these differences are capable of so many different interpretations that they are not worth pursuing.

No doubt it is sociologically significant that, for instance, the proportion of the self-employed males differs significantly among those who stated their religious affiliation. It ranges from 12.1 percent among 'other denominations' and 12.6 percent among Presbyterians, down to just over 8 percent of adherents of the Church of Ireland. But such differences have no real bearing on labour market structure, however interesting they will be to students of the local cultures of different parts of the Province.

As far as projecting these structures into the future is concerned we shall, in Chapter 6, comment on self-employment at the local level; similarly, we shall examine the educational achievements and current educational participation in much greater detail. As far as this overview is concerned, we can only summarize it by saying that the position of the Catholic community does not appear particularly unfavourable. There are local differences in structure, connected largely with the nature of the economy, but they do not differentially impinge upon the denominations.

Development and Incidence of Unemployment

We now turn to the crucial second stage of employment analysis: how did unemployment develop in Northern Ireland, and how differentiated is its spatial incidence? In the Additional Tables A3.1–A3.4 we show the overall development of the civilian labour force, and those in employment, which forms the background to the present analysis. For the central exercise, however, the relevant data are summarized in Table 5.3. We note that the main sources of information now change. The 1981 Census will only be useful in analysing the religious composition of the unemployed; evolution over time will have to be derived from statistics compiled for travel-to-work areas by the Department of Economic Development. Note that the best way of comparing DED statistics with the Census is to assume that 'estimated employees' means economically active males and females, including unemployed, less self-employed. For 1981 the Census would have given 572,000 on that definition, the DED statistics for April 1981 576,000; so we are justified in accepting this identity. Since self-employed persons cannot claim unemployment benefit, they have to be left out of this section of our calculation. (We note that in Table 5.4, where the working population of Northern Ireland and the UK are compared, the self-employed are included in both cases, and this larger definition will also have to be used for the analysis of industrial structure.)

To maintain comparability we have grouped TTWAs in this section into our four area groups, although, as has been previously stated, these do not conform exactly to local government district groupings. However, it is possible to maintain a measure of consistency over time for the main comparisons. The boundaries of TTWAs were changed in 1984, and therefore the latest figures are not comparable in detail with some earlier series, though the trends are not concealed by these changes.

Table 5.3 *Overview of development of unemployment, 1976–85, by area group*

	Unemployed			Estimated employees[2]			Unemployment rates (%)		
Area group[1]	Male	Female	Total	Male	Female	Total	Male	Female	Total
I Belfast									
1976	15,265	7,638	22,903	177,410	118,430	295,840	8.6	6.4	7.7
1979	20,571	11,331	31,902	179,113	127,458	306,571	11.5	8.9	10.4
1981	29,772	14,566	44,338	179,113	127,458	306,571	16.6	11.4	14.5
1983	36,851	14,439	51,290	179,113	127,458	306,571	20.6	11.3	16.7
1985	43,365	17,593	60,958	192,616	147,037	339,653	22.5	12.0	17.9
II Northern									
1976	4,160	2,286	6,446	40,750	26,000	66,750	10.2	8.8	9.0
1979	5,922	3,395	9,317	42,945	30,113	73,058	13.8	11.3	12.8
1981	10,484	4,066	14,550	42,945	30,113	73,058	24.4	13.5	19.9
1983	12,183	4,276	16,459	42,945	30,113	73,058	28.4	14.2	22.5
1985	9,099	3,341	12,440	33,105	25,671	58,776	27.5	13.0	21.2
III Southern									
1976	8,960	3,820	12,780	57,510	40,310	97,820	15.6	9.5	13.1
1979	10,758	5,816	16,574	62,894	45,111	108,005	17.1	12.9	15.3
1981	17,426	6,837	24,263	62,894	45,111	108,005	27.7	15.2	22.5
1983	19,058	7,067	26,125	62,894	45,111	108,005	30.3	15.7	24.2
1985	17,789	7,234	25,023	56,128	42,510	98,638	31.7	17.0	25.4

Table 5.3 *continued*

Area group[1]	Unemployed			Estimated employees[2]			Unemployment rates (%)		
	Male	Female	Total	Male	Female	Total	Male	Female	Total
IV Western									
1976	8,677	3,148	11,825	44,240	27,120	71,360	19.6	11.6	16.6
1979	9,487	4,322	13,809	48,643	31,592	80,235	19.5	13.7	17.2
1981	15,658	5,024	20,682	48,643	31,592	80,235	32.2	15.9	25.8
1983	17,208	5,291	22,499	48,643	31,592	80,235	35.4	16.7	28.0
1985	18,690	5,181	23,871	50,005	34,006	84,011	37.4	15.2	28.4
Northern Ireland									
1976	37,062	16,892	53,954	319,910	211,860	531,770	11.6	8.0	10.1
1979	46,738	24,864	71,602	333,595	234,274	567,869	14.0	10.6	12.6
1981	73,340	29,450	101,522	329,050	247,100	576,150	22.3	11.9	17.6
1983	83,440	30,229	113,669	316,000	242,000	558,000	26.4	12.5	20.9
1985	88,924	32,757	121,681	331,900	249,200	581,100	26.8	13.1	20.9

[1] For definition of area groups see Appendix A. Travel-to-work area boundaries were changed between 1983 and 1985, and figures are not therefore strictly comparable.

[2] Estimated employees (base) not changed annually, so percentages are not accurate.

Source: Department of Economic Development Press Notices, 1976–85

Table 5.3 shows that in the period 1976 to 1985 the estimated number of employees rose from over 530,000 to 581,000, a rise of just under 10 percent (see also Figure 18). This is, of course, significant: one could say that a large part of total unemployment growth was due to an increase in the size of the potential labour force, and that all that happened was that the economy failed to keep pace. However, in Chapter 4 it has already been demonstrated, in respect of males only, that this is too simple a view of the matter.

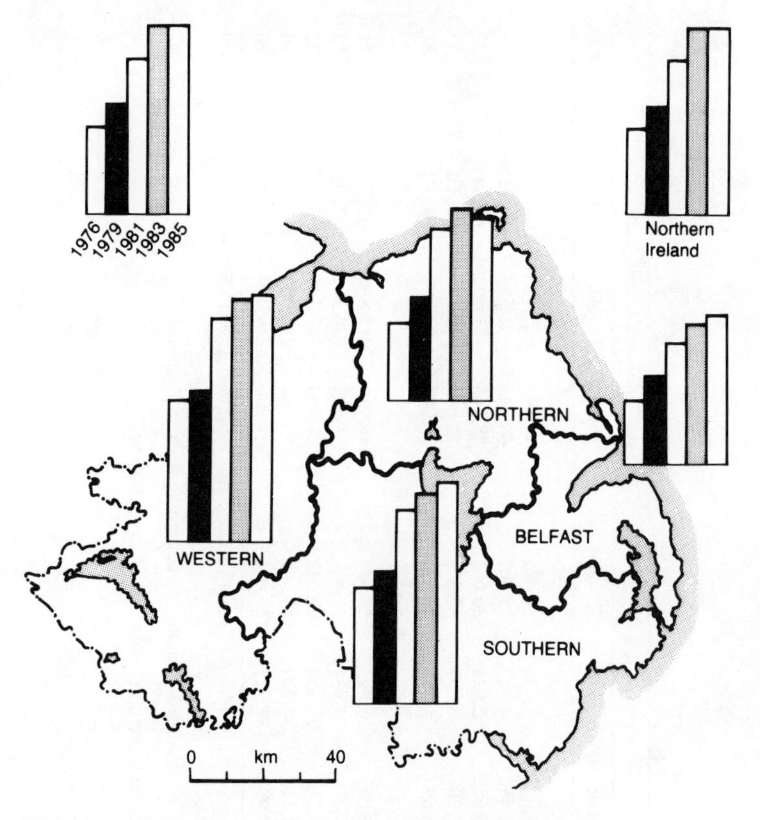

Figure 18 *Overview of development of unemployment since 1976, by area group (see also Table 5.3)*

Source: Department of Economic Development Press Notices 1976–85

The increase in the (differently counted) 'total working population' in Northern Ireland between 1973 and 1983 was nearly 7 percent (see Table 5.4), whereas in the UK as a whole it was only about 3.5 percent. However, the rise in unemployment was very much greater than can be accounted for by this demographic change.

Table 5.4 *Distribution of total working population, employed and unemployed, United Kingdom and Northern Ireland, 1973–83 (thousands)*

	1973			1975			Percentage change	1983			Percentage change
	Male	Female	Total	Male	Female	Total	1973–75	Male	Female	Total	1975–83
Total working population											
United Kingdom	16,245	9,368	25,613	16,162	9,715	25,877	+1.03	16,165	10,611	26,776	+3.47
Northern Ireland	387	199	586	396	220	616	+5.12	403	256	659	+6.98
Unemployed											
United Kingdom	476	81	557	698	140	838	+50.5	2,145	839	2,984	+256.10
Northern Ireland	27	7	34	26	9	35	+3.0	83	30	113	+222.90
Total employees in employment											
United Kingdom	13,773	8,891	22,664	13,536	9,174	22,710	+0.20	11,982	9,228	21,210	−6.61
Northern Ireland	295	186	481	296	201	497	+3.33	250	216	466	−6.24

Sources: For UK: *Annual Abstract of Statistics 1985* (Central Statistical Office, 1986).
For Northern Ireland: *Annual Abstract of Statistics* nos 2 and 3

Furthermore, most of this overall increase in the total working population, whichever measure we take, occurs among women workers: the male labour force changes very little either in Northern Ireland or in the UK as a whole, in the decade under review, whereas the female labour force increases sharply. Yet, as both Tables 5.3 and 5.4 show, it is male unemployment which rises, in the UK by a factor of four between 1973 and 1983, and in Northern Ireland by a factor of three, thus accounting for the greatest part of the increase in unemployment. (In percentage terms, female unemployment rises faster but the numbers involved were so small at the start of the period that this is something of a statistical illusion.) The result is that for Northern Ireland overall unemployment figures (Table 5.3) more than double, from 54,000 in 1976 to 122,000 in 1985, but 52,000 of that increase is among males and only 16,000 among females.

The increase in female workers is generally held to be a response to market demand, due to changing technology and changing industrial composition, and we shall investigate this further in Chapter 6. It is also claimed (at least by men) that women are 'taking men's jobs', and this is then demonstrated by reference to the fact that in the decade the number of women in the workforce increased by almost the same amount as the rise in male unemployment. As we shall see, there is little substance in this facile inference. However, it is true that employers do, whenever they can, substitute (mainly part-time) women workers for full-time men. In the UK as a whole the rise in male unemployment is also only a little larger than the increase in women workers, and similar claims are made there, with as little justification, for the inference of a direct causal linkage.

In much of the rest of this report we will concentrate our analysis again on the men, as we did when we looked at future entrants to the labour market. This is not to underestimate that unemployment for women can mean just as much hardship, especially for single, widowed and divorced women. Given the low overall household incomes, the contribution made by a second earner can be of crucial importance in determining living standards. If, as is so often the case in Northern Ireland, a large proportion of adult males are long-term unemployed, the loss by a woman even of a relatively badly paid factory job can mean all the difference between meagre survival and stark poverty. Nevertheless, given the fact that Northern Ireland is, compared with Great Britain, still rather a traditional society, where a higher proportion of all women spend a longer period of their lives in child-rearing, female full-time employment plays a somewhat less significant role in the economy than it does in Britain, especially the southern half of the country. Because of fewer women having insurance entitlements, however, the true figure of women 'seeking work' (in Census terms) should be rather greater than the unemployment statistics indicate. If we look at the employment status statistics of the 1981 Census economic activity tables (Table 5.5), however, we find that of the 346,000 males who were economically active and not self-employed, 22 percent were seeking work, and that figure agrees very well with the claimant-based unemployment in Table 5.6. But for women, the same Census calculation gives us 9.4 percent seeking work.

It is always difficult to reconcile the various statistics of unemployed women: those registered for benefit, and those recorded as 'seeking work' in the Census. In the former case, entitlement to benefit is the criterion (and this implies the willingness to work). In the latter, a woman is not allowed to record herself *both* as a housewife (that is, economically inactive) *and* as seeking work, though she may well wish to join, or rejoin, the labour force. She may also have a part-time job, and this will mean that she appears in the statistics of women in employment, but not in the figures of persons registered for benefit, or in the Census classification of 'seeking work'.

When we look at the claimant-based tables, we find 13.1 percent (penultimate column of Table 5.3) of the workforce (that is, estimated employees) unemployed among the women. The difference arises, as inspection of the two tables shows, from the fact that the Census workforce, in the case of women, is larger than the official (insurance-based) count, so that the same number of unemployed women (29,000 both in the Census and in the DED statistics) constitute a higher proportion of job-seekers in the first case than in the second.

There are other small discrepancies in total numbers in employment (employed or self-employed) and unemployed when one compares the male labour force in the Census and the annual figures from the DED. In contrast to the figures for women, the annual counts show *fewer* men at work than the decennial Census at roughly comparable dates. Neither these (relatively small) differentials, nor the fact that the deviation goes in opposite directions for men and women, can be easily explained. Since in what follows we have to use mainly Census statistics (to ascertain the denominational breakdown of the labour force), consistency will be maintained, given the limitations of the Census in general.

It should be noted in passing that for many years it has been a commonplace that official unemployment statistics understated unemployment in the UK compared with other industrialized countries, because British figures were based on benefit claimants, as opposed to the majority of countries which used Census or survey

Table 5.5 *Employment and unemployment in Northern Ireland, males and females, 1981, by stated religion, by area group (thousands)*

Area group[1]	Economically active		Self-employed		Potential employees		Unemployed		Unemployed as % of potential employees	
	Male	Female	Male	Female	Male	Female	Male	Female	Male	Female
All denominations										
I Belfast	181.9[2]	117.9	15.0	2.2	166.9	115.8	29.3	13.2	17.6	11.4
II Northern	65.1	34.7	10.3	0.9	54.9	33.8	11.9	4.1	21.8	12.1
III Southern	89.8	48.6	16.2	1.3	73.6	47.3	19.8	7.3	26.9	15.4
IV Western	61.5	30.1	10.2	0.7	51.3	29.4	16.2	4.6	31.7	15.7
Northern Ireland	398.4	231.4	52.6	5.1	345.8	226.3	76.3	29.2	22.1	12.9
Roman Catholics										
I Belfast	28.5	19.4	2.1	0.3	26.4	19.2	7.8	3.2	29.5	16.6
II Northern	13.6	7.3	2.3	0.2	11.3	7.1	3.8	1.1	32.9	15.4
III Southern	33.0	18.0	5.2	0.4	27.8	17.7	10.2	3.3	36.7	18.9
IV Western	27.7	14.2	3.9	0.3	23.9	14.0	9.3	2.5	39.4	17.9
Northern Ireland	103.0	59.0	13.5	1.1	89.3	57.9	31.1	10.1	34.8	17.4
All others										
I Belfast	153.4	98.6	13.0	2.0	140.4	96.7	20.5	10.0	14.6	10.4
II Northern	51.6	27.5	8.8	0.7	42.8	26.7	8.2	3.0	19.2	11.2
III Southern	56.8	30.5	11.0	0.9	45.8	29.6	9.6	4.0	20.9	13.4
IV Western	33.8	15.9	6.3	0.4	27.5	15.5	6.9	2.2	25.0	13.7
Northern Ireland	295.6	172.4	39.1	4.0	256.5	168.4	45.2	19.1	17.6	11.3

[1] For definition of area groups see Appendix A.

[2] Numbers may not always total due to rounding.

Source: Own calculations based on Northern Ireland Census 1981, *Economic Activity Report*, Table 9 (unpublished information)

Table 5.6 *Employment and unemployment structure of Northern Ireland, by religion and by area group, 1981, males only*

Area group[1]	All denominations	Roman Catholic no.	Roman Catholic %[2]	All other no.	All other %	Not stated no.	Not stated %
I Belfast							
Total 16+	232,235	36,481	15.7	154,969	66.7	40,785	17.6
Inactive	50,349	7,975	15.8	35,187	69.9	7,187	14.3
Out of employment	28,291	7,786	27.5	14,348	50.7	6,157	21.8
In employment	153,595	20,720	13.5	105,434	68.6	27,441	17.9
II Northern							
Total 16+	83,561	17,819	21.3	52,451	62.8	13,291	15.9
Inactive	18,413	4,222	22.9	11,787	64.0	2,404	13.1
Out of employment	11,948	3,722	31.2	5,566	46.6	2,660	22.3
In employment	53,200	9,875	18.6	35,098	66.0	8,287	15.6
III Southern							
Total 16+	116,597	43,358	37.2	50,971	43.7	22,268	19.1
Inactive	26,786	10,320	38.5	11,864	44.3	4,602	17.2
Out of employment	19,790	10,218	51.6	4,488	22.7	5,084	25.7
In employment	70,021	22,820	32.6	34,619	49.4	12,582	18.0
IV Western							
Total 16+	80,650	36,939	45.8	29,217	36.2	14,494	18.0
Inactive	19,111	9,238	48.3	6,723	35.2	3,150	16.5
Out of employment	16,240	9,368	57.7	3,174	19.5	3,698	22.8
In employment	45,299	18,333	40.5	19,320	42.7	7,646	16.9
Northern Ireland							
Total 16+	513,043	134,597	26.2	287,608	56.1	90,838	17.7
Inactive	114,659	31,755	27.7	65,561	57.2	17,343	15.1
Out of employment	76,269	31,094	40.8	27,576	36.2	17,599	23.1
In employment	322,115	71,748	22.3	194,471	60.4	55,896	17.4

[1] For definition of area groups see Appendix A.

[2] Percentage of all denominations; percentages may not add up to 100 due to rounding.

Source: Own calculations based on Northern Ireland Census 1981 (unpublished information)

sources. However, in 1981 the situation was reversed and the Census showed fewer unemployed than the labour statistics. The same seems to have been true in Northern Ireland.

The outcome of these considerations is that we may accept both Census and non-Census statistics of unemployment with a certain degree of confidence, always with the proviso that the true number of men seeking work may be higher than that recorded by the DED. We shall analyse the female unemployment rates less thoroughly than the male, because of the small contribution of women to total income, and not because their problems are less important for them and their families.

In Table 5.3 we look at the four area groups in greater detail. It will be noted that the denominator group for the unemployment calculation changes in 1976 and 1979 at the local level, but then not again until the 1985 figures: this is because the DED could not give an accurate estimate of the economically active labour force by individual travel-to-work areas which, as we have said, do not correspond to districts, and in any case annual estimates of population age and sex composition at local level are notoriously unreliable. So we simply note from Table 5.3 that the biggest increase recorded relates to Belfast, and is larger for women than for men. Part of this is due, however, to a transfer of working populations between the reorganized TTWAs (see Additional Tables for the distribution of employed and unemployed at TTWA level).

Since, however, the unemployed were always counted in relation to the appropriate assumed employed population, the percentage figures in the last three columns of Table 5.3 may be taken as reflecting the local differentials accurately enough, as can the changes over time. If anything, the Belfast figures might be held to understate the true unemployment position because of the very sharp upward revision of the estimated employees. By the same token, the Western area group calculation may somewhat overstate the position because the upward revision of the male employees seems to be rather on the low side.

These reservations apart, the picture is clear enough: unemployment rose linearly over the years shown, and the unemployment gradient was the same as that to which we have become accustomed, culminating in the figure of 37.4 percent unemployment for males in the Western area group in 1985. At these levels, a percentage point or two either way because of some argument about boundaries really does not seem to matter. The figures in the last column may be taken at their face value. We note, however, for what it is worth, that unemployment rises have rather tended to level off in the last two years under review, and if this relative stabilization is real, we shall have to explain why this may be so. We shall investigate this point when we look at the industrial structure of the country.

Unemployment and Religious Denomination

In Table 5.5 we look at the differences in unemployment between the religious groups, by area. We note that the overall (Census) figure of unemployment is about 105,000, made up of 76,000 men and 29,000 women. These totals constitute 22 percent of all men economically active and not self-employed, and 13 percent of all women on the same definition. If we look at the declared Roman Catholics only, we find 34.8 percent of the men unemployed, and 17.4 percent of the women. In the case of men, this figure is 58 percent higher than the total for all denominations, and in the case of women, 35 percent higher. If we now exclude the Catholics and take all the rest of the workforce, including those who did not state their religious affiliation, the male percentage is 17.6 and the female 11.3.

Thus, male unemployment is nearly twice as high for Catholics as for all men, and more than 54 percent higher for women. In Additional Table A5.1 we have reproduced figures for all stated religions. From these detailed statistics we derive the information that for all men stated as Presbyterians, Church of Ireland, and Methodists, the unemployment rate was 14.4 percent. This would make Catholic male unemployment 140 percent higher than Protestant. For women the corresponding percentage would be just under 10, so that the Catholic rate is about 75 percent higher than the Protestant rate. Clearly, then, the 'not stated' category must include a higher proportion of unemployed than that for Protestants: 27.1 percent of the 'not stated' males were unemployed, and 16.1 percent of the women. So the conclusion of this would be that the 'not stateds' unemployment pattern was much nearer that of the Catholics. It is clearly tempting to divide up the 'not stateds' in some way so that the sum of proportionate unemployment for Catholics and Protestants comes to the respective totals. Mathematically this is not difficult: unfortunately the result is not usable. Only if we allocated 45 percent of 'not stateds' to the Catholics, and then applied the Catholic unemployment ratio, and 55 percent to the Protestants, with *their* ratio, would we get the total of all 'not stated' unemployed. This would be an absurd proceeding because we have so many other Census classifications by religion, and if in each case we divided the total of 'not stateds' so that any subset agreed with the declared proportions among the Roman Catholics, we would get a different figure each time. To put this difficulty in another way, if we took what we believe to be the overall proportion of Catholics in Northern Ireland, as stated in our religion report (Eversley and Herr, 1985), we would have a fixed way of allocating the 'not stateds': 55 percent would have to be Catholic, and 45 percent Protestant and 'other'. But if we did this in the present case, then the Catholic proportion of unemployed applied to over 40,000 'not stateds', and the Protestant proportion applied to the other 33,000, would lead to far more unemployed than we actually have.

So we shall not pursue this method in our detailed tables: we shall compare the stated Catholic ratios with those for all other persons, including 'not stated'. As we have seen, the difference is sufficiently large to have serious policy implications. Whether 'true' Catholic unemployment is twice as high as that for Protestants, or 65 percent higher, becomes, in the context of this report, a little academic.

In Table 5.5 we also analyse unemployment by religion, by area. The base figure here consists of the economically active men and women recorded in the 1981 Census (disregarding self-employment). As before, the 'not stated' respondents are not allocated between religious groups. Even if we accepted the assumptions about the correct division of this group at a national level (and various figures have been suggested: Hepburn, 1982; Compton, 1981; Fair Employment Agency, 1978; Miller and Osborne, 1983), it would be impossible to use the same proportions at area, let alone at district level. Thus Table 5.5 must be regarded as indicating orders of magnitude of differences, not absolute figures.

Overall, (Census) unemployed Roman Catholics, 41,200, account for 6.5 percent of all economically active persons, but they constitute just over a quarter of all stated Catholic economically active people. As we have learned to expect, this proportion varies along the usual gradient from centre to periphery. The figure for Catholic unemployment is lowest where Catholics also form the smallest part of the labour force, especially in the Belfast suburban areas. (For a full breakdown of unemployed, see the Additional Tables.) Of the 105,000 (Census) unemployed in 1981, the 41,200 Catholics constituted 39 percent – exactly the same as their total share of the population as computed by us (Eversley and Herr, 1985). This overall proportion is, however, heavily weighted by persons under working age. The correct comparison therefore would be between persons over 15: among the adults, we calculated 36.4 percent to be Catholics. The unemployed male Catholics came to 40.7 percent of all males, and the women to 34.6 percent of all females.

The differences at district and area level are much larger than for the country as a whole. Not all the statistics have been reproduced in tables and figures in this volume, but the essence of the matter is clear from an inspection of Table 5.6 and Figure 19. If we take the Northern Ireland proportion of unemployed males as our benchmark (22 percent in 1981), locally this varies from 8.7 percent in North Down to 39.7 percent in Strabane. For women the benchmark is 12.9 percent, and locally this varies from 7.2 percent in Castlereagh to 21.2 percent in Cookstown. If we turn to the Roman Catholic population, the overall male rate was 34.8 percent, but locally this varies from 9.7 percent in Castlereagh to 53.6 percent in Cookstown. For women the overall Catholic rate was 17.4 percent, and ranged from 7 percent in Castlereagh to 27.3 percent in Cookstown. In contrast, the

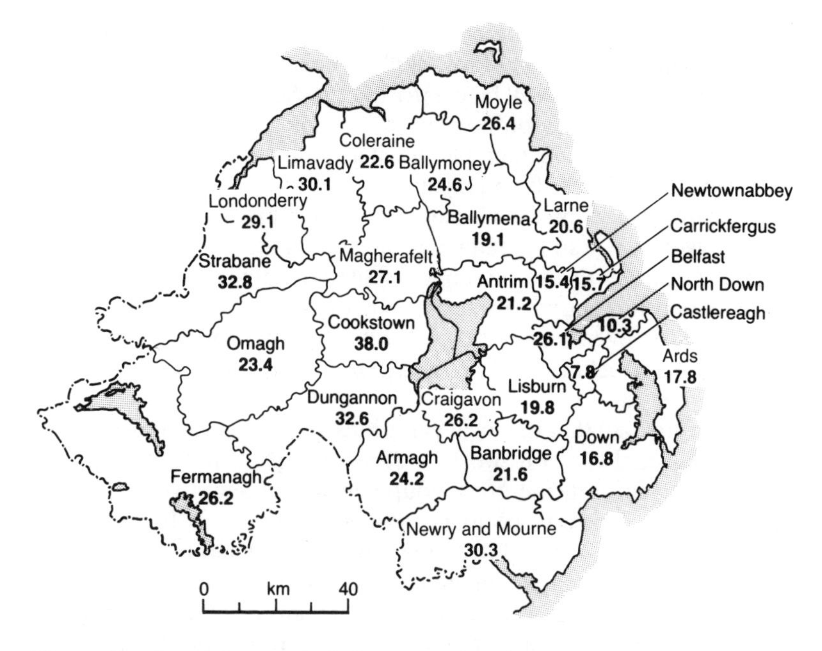

Figure 19 *Stated Roman Catholics unemployed as a percentage of economically active, by district, 1981 (see also Table 5.5)*

Source: Own calculations based on Northern Ireland Census 1981, *Economic Activity Report*, Table 9 (unpublished information)

'all other plus not stated' category shows an average of 17.3 percent for men and 11.3 percent for women, with local values of 34.3 percent in Strabane for men, and 18.2 percent for women in Cookstown, down to 9.9 percent for men in Castlereagh and 7.3 percent for women in the same district.

The large variations in the size of the difference between the denominations is partly a matter of the local denominational mix, but even more so should be seen as the outcome of a different socio-economic structure: the middle-class Belfast suburbs cannot really be compared with peripheral part agricultural, part industrial areas.

If an area has a very high percentage of Catholic adults, then the Catholic unemployment rate will be much nearer the local average. The rate for the 'others' will, correspondingly, be indeed lower than that for Catholics, but not by a large margin. In areas where Catholics are a small minority, but where the local socio-economic structure is biased towards the managerial and professional groups and other white-collar workers, the differences will not be pronounced. However, in areas where the Catholics are both a relatively small minority, and where the socio-economic structure shows a much higher proportion of low-skill groups, the difference between Catholics and others will be *greater* than is the case nationally.

If we take the 2:1 proportion of Catholic male unemployment percentage to all male unemployment as standard, then we show it to be twice as high in Ards and Lisburn. If we pulled out the 'Protestants only' categories from the Census tables, then the result would be, as in our overview, a still greater contrast, but in the same directions. In those cases where the unemployment percentage for men in the 'others and not stated' group is almost exactly the same as for the Catholics or even higher (as in Strabane), there will be both few Protestants altogether and a very high proportion of Catholics among the 'not stated'.

Given the fact that the local variations are exaggerated because of residential and travel-to-work patterns, we give, in Tables 5.5 and 5.6, the position in the area groupings. These again show nothing unusual; the centre to periphery gradient appears as before. Area group I (Belfast) has the lowest overall unemployment, as well as the lowest Catholic unemployment for men, though women's unemployment is marginally lower in the Northern group. 'Others' and 'not stateds' exhibit the same gradient from centre to periphery. As we would expect, Catholic unemployment is nearest to the total percentage in the Western area, but furthest away from 'others' in the Belfast group. Catholic women's unemployment is lower overall than Catholic men's, and the interdenominational differences are not as large. As we shall see, this can be explained because of the concentration of Catholic women in the low-paid public sector jobs. The unemployment table therefore does not reflect the differences in household income sufficiently.

The Belfast group accounted, in 1981, for 45 percent of the economically active males, and 48 percent of the potential employees. For Catholics, 27.6 percent of economically active males were in the Belfast group, and 29.5 percent of the potential employees. If we analyse the distribution of unemployment, 38 percent of all unemployed men were in Greater Belfast, but only 25 percent of the Catholic male unemployed. Conversely, for the 'others' and 'not stated' groups, 52 percent of their economically active people were in Belfast, 54.7 percent of their potential employees, and 45.5 percent of their unemployed.

The conclusion one would draw from that would be that Roman Catholics had a relatively better chance of being employed in Belfast than the others. The distinction is more apparent than real, but we shall have to bear it in mind when we look at sectors of the labour market. The position is clearly somewhat distorted by the middle-class suburbs with their high proportions of salaried and professional Catholic workers. If we look at Belfast district on its own, it had 19.6 percent of the economically active male population, 21.5 percent of the potential employees, and 22.1 percent of the unemployed. But declared Catholics in Belfast constituted 18.9 percent of their national total, 20.6 percent of their potential employed and 19.6 percent of their unemployed. The main advantage therefore must have lain in the outer districts, and this is confirmed by looking at figures for these.

For women, the position is somewhat different: 51 percent of the economically active (and potential employees) are in Greater Belfast, but only 45.3 percent of their unemployed. So for the whole population the proportions are similar for men and women, but for Catholics 33 percent of all potential employees are in Belfast, and 31.4 percent of their unemployed. So the women's position in that labour market is more unfavourable than the men's.

When we turn to the Catholic women in Belfast district alone, it can be seen that they form nearly 30 percent of the country's female Catholic potential employees, but less than 25 percent of the unemployed. This suggests that their chances are marginally better in inner Belfast than in the outer areas. This again needs to be remembered later: the low-paid jobs are concentrated on the city, the higher-paid ones on the outer areas. Therefore the men fare relatively better in Greater Belfast than in the city, and for the women it is the other way round.

Area group IV (Western) had 14.8 percent of potential male employees, and 21.3 percent of unemployed; for Catholics the proportions are 26.7 percent and 30.1 percent. For the 'others', 10.8 percent of the potential employees are in that group, and 15.2 percent of their unemployed. So in a narrow sense one could say that the Catholic 'excess' unemployment is less than that of the 'others' – but this is fairly meaningless when the absolute Catholic male rate of unemployment is nearly 40 percent, and that of 'others' 25 percent.

If we do not pursue the analysis of these tables any further, it is not because they do not give us a great many clues to local variability. Rather, the proportions will in general be in accordance with a pattern already well established, and until we analyse the industrial structure the differences must remain merely suggestive.

In Table 5.6 we look at the matter in a slightly different way: we analyse the employment status of Catholic men, 'all others' and 'not stateds' at the level of the whole country, and in the four area groups, as a proportion of all persons in each status group. Thus for Northern Ireland the stated Catholics were 26.2 percent of the over 16s, the 'others' 56.1 percent and the 'not stateds' 17.7 percent. Levels of inactivity were about equal, as we have already demonstrated. But of those out of employment, 40.8 percent were Roman Catholics, 36.2 percent were 'others' and 23.1 percent were 'not stateds'. Of those in employment (here including the self-employed, which is correct as we are not measuring the percentages of those technically unemployed), 22.3 percent were Catholics, 60.4 percent Protestants and 'others', and 17.4 percent 'not stated'. Thus, the proportion of Catholics who were unemployed was nearly twice as high as that of Catholics as a part of the working population; for 'others', in contrast, the proportion of these in the population at work was nearly twice as high as it was among the unemployed.

At the area group level, these differences are sharpened. We have the usual centre to periphery gradient. In the wider Belfast region, Catholics formed 27.5 percent of all unemployed and 13.5 percent of the employed. In the Northern area they were 31.2 percent of the unemployed and 18.6 percent of the employed. In the South the figures were 51.6 percent and 32.6 percent, and in the West 57.7 percent and 40.5 percent. In other words, the higher the proportion of Catholics aged over 16 years, the higher the proportion they formed of those at work, but also of the unemployed.

In contrast, for the 'others', unemployment rates were nearest the total age group rate, and the proportion of those at work, in the Belfast region. In fact, in this area the proportion of 'others' in work was almost the same as their share of the adult population. Then, as we move away, the differential shows up: everywhere else the proportion of men working in this category was higher than the population proportion, and the unemployment rate much lower, until we get to the West, where the share of those in work was twice as high as their share of unemployment, and 18 percent higher than their share of population.

Looking at the last column of Table 5.6 we find that the 'not stated' category is roughly the same proportion in all four area groups, and its share of those out of employment is also about the same, but larger than its share of population. The proportion of men at work is again very similar in all areas, and not too far away from their population share. (This table alone shows how dangerous it would be to divide the 'not stated' category into denominations by some rule of thumb, however mathematically arrived at.)

The 1981 Labour Market in Perspective

Most adults below the age of retirement were in work in Northern Ireland in 1981. Among men, those at work were the great majority, but rather less than half the women. Unemployment was growing. More potential workers were entering the labour market than were retiring from the workforce. A high proportion of all women who were at work were part-timers. A high proportion of all workers, but especially women, were in relatively low-paid jobs and had little prospect of promotion.

Employment was unequally distributed. In Greater Belfast, the overall position was no worse than in the disadvantaged areas in Great Britain: Glasgow, Merseyside, the North East. Outside Belfast, conditions were worse than in any other region of the UK, and that too was the overall position of Northern Ireland.

So far we have identified the nature of the inequalities in the labour market, but not the underlying causes. Clearly, the more remote the district from the Belfast centre, the worse its position. The more rural and thinly populated the area, the fewer the chances it offered. Manufacturing industry declined everywhere, and the loss of employment was only partially and unevenly compensated by the rise of service occupations. The public sector played an increasingly significant role, and more so where alternative employment was disappearing fastest. Private service jobs were not expanding in areas of low purchasing power.

Women's jobs expanded where men's jobs contracted. Many of the female occupations were characterized by much part-time working, low pay and poor prospects. This applied especially in the largest public sectors, like education and health. As the recession deepened, the public sector ceased to expand.

Within each labour market the Roman Catholic community was more disadvantaged, had more men and women unemployed, and did not have a fair share of the better types of jobs, managerial and professional. Only in the agricultural sector, with much self-employment, did Catholics hold their own. Yet the differences were not particularly marked within each subsection of the labour market. The proportion of Catholics was highest in the more rural areas outside Belfast, and it was there that unemployment and low-grade jobs were most common. Only within the Belfast area was their relative disadvantage manifest at the local level. Derry, the most Catholic urban area, bore the brunt of the industrial decline of the west, and exhibited all the signs of a failing economy in every respect.

The Catholic disadvantage then is twofold: relatively more of them live in declining areas, especially on the periphery of the country; and within the slightly more prosperous areas, at or near the centre, they bear the brunt of rising unemployment. Only in those areas of Greater Belfast, where the middle-class professional and managerial populations are concentrated and where most of the higher-echelon public servants live, does the Catholic disadvantage become less apparent.

All this has been well known for a long time, though we have added a more detailed analysis for particular areas, industries and occupations. We still have to ask some more questions, however, before we try to allocate the reasons for the state of affairs to different kinds of cause. How mobile are the Catholics within the labour market? How far do they have, at least in theory, access to possible jobs? In an earlier chapter we dealt with geographical mobility and decided that, overall, people in Northern Ireland seemed to have moved not much less than people in Great Britain, at least in the decade 1971–81; taking an even longer perspective, emigration to the rest of the UK and overseas has clearly been very important.

But this is not the same thing as saying that the Catholics have been able, or are able now, to move where the best job opportunities are. This will be a matter of housing availability, and of their feelings of personal security even if they are adequately housed (as in the improved wards of Belfast city and in the suburban NIHE estates). Additionally, they may not be able to reach workplaces (supposing they were offered) because of demarcation lines not visible to the outsider. In this report we cannot deal with these aspects of the matter. The last major field we have to address is the question of education and training. Is the (younger) Catholic population mobile in the sense that its school, college and university opportunites adequately fit it for such openings as the labour market can offer, especially of the better kind? Clearly if, as has been claimed, such people are not suitably qualified, employers would appear to be justified in discriminating against them, up to a point. But is this really so? That at least it should be possible to establish.

6

Development of the Industrial Structure

We now turn to the analysis of the core of the *demand* side of the labour market equation. The change in industrial structure of the country has been the subject of a number of monographs and is still under intensive investigation (Bradley et al., 1984; Coopers and Lybrand, 1982; Northern Ireland Economic Council Reports nos 38, 45, 46, 51). However, most of this research concentrates on the structure at the level of the Province as a whole and does not relate to travel-to-work areas, let alone districts. Nor has the religious composition of the labour force been analysed in terms of the changing industrial structure.

Our sources of information are again twofold. First, the 1981 Census is our only source about the religious composition of the labour force by industry at the local level. Secondly, there are annual data produced by the Department of Economic Development, which until 1981 provided information on the basis of the old 27 standard industrial classifications, and has since then used the new 75 industry divisions. The changeover makes analysis of time trends a little difficult. We have tried to solve the problem by providing broad industrial groupings which are roughly comparable (see Tables 6.4, 6.6 and 6.7; see also Additional Tables A6.4 and A6.5(b)). Since the classification is supposed to be applicable to the whole of the UK, a great many categories appear in the full data which have no importance at local level in particular, and others about which we would like to know more have not been subdivided. In some cases we have relied on special extraction of subdivisions (Table 7.3).

First, however, we need to give a brief overview of two sectors of the total labour market – self-employment and public sector employment – which will in one form or another be a part of every occupational and industrial analysis in the later parts of this report.

Self-employment

In one sense, self-employment has to be taken out of this analysis of the labour market: for a long time now, self-employment has been fairly stable (see Additional Table A6.7). Well over half the total self-employed are farmers (industry division group I in Table 6.2) and their number has hardly declined in a decade, though no doubt some very small holdings may disappear through amalgamation.

In manufacturing (industry division group II in Table 6.2) there has been a rising trend, predictably concentrated on such products as furniture and toys. This is not the place to discuss the success of the many governmental schemes to encourage small business: the fashion has persisted for some time and the gain of 600 self-employed people (almost all of them men) is not negligible, though it is somewhat dwarfed by the overall increase in the loss from manufacturing industry. There has been some loss in the retailing sector – again in line with the experience of most countries, with no countervailing increase in other forms of service occupations.

Table 6.1 *Self-employment, 1981, by area group*

Area group[1]	Total economically active population	Total self-employed	Self-employed as % of all civil economically active
I Belfast	299,829	17,215	5.7
II Northern	99,869	12,012	12.0
III Southern	138,403	17,545	12.7
IV Western	91,658	10,911	11.9
Northern Ireland	629,759	57,683	9.1

[1] For definition of area groups see Appendix A.

Source: Own calculations based on Northern Ireland Census 1981, *Economic Activity Report*, Table 4

Table 6.2 *Evolution of self-employed, males only, by new industry division groups, 1974–83*

Industry division groups[1]		1974	1978	1983
I	01–17	43,140	42,270	39,430
II	21–49	1,490	1,840	2,140
III	50–79	23,210	22,860	22,580
IV	81–85	2,990	2,840	2,730
V	91–99	3,050	2,860	2,720
Northern Ireland		73,880	72,670	69,600

[1] For explanation of groups, see Table 6.4. For industry divisions see Appendix B.

Source: Own calculations based on Northern Ireland *Annual Abstract of Statistics* no. 3, Table 10.4

Geographically, the percentage of the economically active population in self-employment follows the expected pattern: very little in the Belfast area, and a higher proportion in the rural areas (see Figure 20 and Table 6.1). We have not presented any denominational

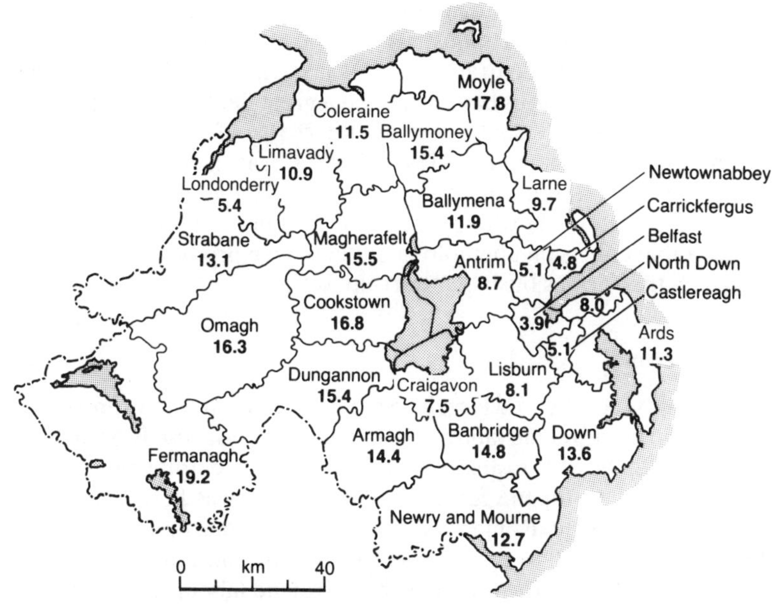

Figure 20 *Self-employment in Northern Ireland, by district, 1981 (see also Table 6.1)*

Source: Own calculations based on Northern Ireland Census 1981, *Economic Activity Report*, Table 4

breakdown of these figures by area, since the observed variations are small, and present no significant pattern. In Catholic areas most farmers are Catholic, and in Protestant areas they are Protestant. Retailing seems to be slightly more likely to be in Protestant hands, but it is difficult to be precise about this. Locally these small businesses may exert a marginal influence on employment patterns, but most of them employ no staff, or only members of their own families. The most one can hope for is that the economy will continue to support the same number of farmers, retailers and providers of professional services as it has done throughout the recession with, perhaps, government schemes helping to set up some new businesses to compensate for those that die, or are absorbed into other enterprises.

If this section sounds dismissive of this sector of the economy, it is not meant to be: it is just that we have to concentrate on the employment prospects of several hundred thousand young people in the next decade, and it is inherently unlikely that the self-employed sector will make any difference to these prospects. This has been the experience also in those regions of Great Britain where mass unemployment has followed industrial decline: the enterprise creation schemes funded by central and local government, the banks and voluntary organizations have attracted a great deal of publicity without ever making any dent in the relentless rise of the proportion of unemployed young people. This is not a reason for withdrawing support for them, but it is appropriate to leave the matter aside in a report dealing mainly with past trends. In other areas of the UK no measurable effects on unemployment had been produced by the mid 1980s by the various business 'start-up' schemes.

Public Sector

The public sector figured largely in the investigations of the prospects of the Northern Ireland labour market in the 1970s, largely because it appeared to be the fastest expanding area of employment, with secure prospects for the future, and because it was here that many observers expected Roman Catholics to have the best chances of finding jobs, as well as the best prospects of promotion.

If we assume that the total Northern Ireland labour market has to cater for about 650,000 people, of whom 500,000 are employees (see overview in Table 6.3 and Additional Table A5.1), the public sector, with about 200,000 jobs, is clearly of crucial importance. In 1976 when there were 578,000 employed people in Northern Ireland, 193,000 of them worked in the public sector, exactly one-third. By 1984, when total employment had dropped to 540,000, there were more than 211,000 people in public sector employment, or nearly 40 percent (see also Figure 21). This change is a function of both the reduction in private sector employment, and a slight rise in the public sector during a period when such employment in the rest of the UK was falling steadily.

The rough breakdown in Table 6.3 does not give many clues as to why the different sections of public service, as recorded by Northern Ireland official statistics, should contribute differentially to the observed growth. Using the more detailed occupational and industrial breakdowns provided partly by the Census and partly by government offices responsible for monitoring economic development (see Additional Table A6.6), we find, not surprisingly, that a large part of the additional employment occurred in the Royal Ulster Constabu-

Table 6.3 *Public sector employment evolution, 1976–84, Northern Ireland only, males, females and totals, by main groups of employers*

	1976			1979			1982			1984		
Employer group[1]	Male	Female	Total	Male	Female	Total	Male	Female	Total	Male	Female	Total
Northern Ireland central government	36,365	12,025	48,290	37,073	13,316	50,389	35,492	13,561	49,053	34,029	13,165	47,194
Bodies under the aegis of NI central government	26,085	74,852	100,937	28,487	85,898	114,385	29,599	91,380	120,979	29,374	93,402	122,776
UK central government	5,945	2,569	8,514	3,737	2,708	6,445	3,549	2,741	6,290	3,487	2,648	6,135
Local government	7,123	1,266	8,389	7,609	1,778	9,387	7,506	2,114	9,620	7,924	2,370	10,294
Public corporations												
NI based	14,017	2,433	16,450	13,913	2,902	16,815	12,909	3,013	15,922	12,886	3,158	16,044
UK based	7,949	2,057	10,006	7,495	2,058	9,753	7,524	2,119	9,643	7,311	1,899	9,210
Total	97,484	95,202	192,686	98,314	108,660	206,974	96,579	114,928	211,507	95,011	116,642	211,653

[1] For full list of bodies in each employer group see Additional Table A6.6(a).

Source: Own calculations based on Northern Ireland *Annual Abstract of Statistics* no. 3, Table 10.7

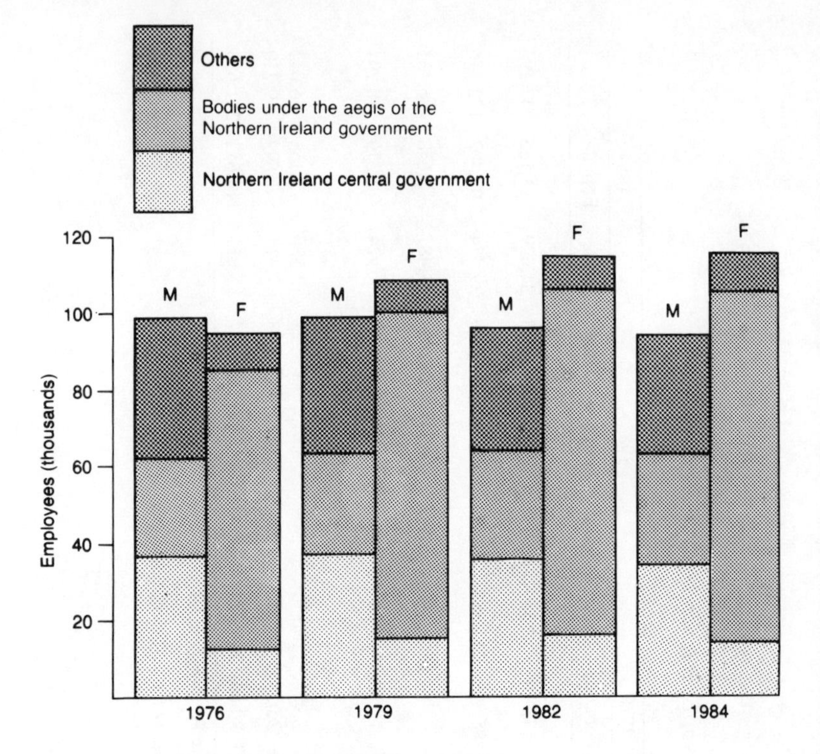

Figure 21 *Public sector employment, 1976–84*

Source: Own calculations based on Northern Ireland *Annual Abstract of Statistics* no. 3, Table 10.7

lary and in the Prison Service. Owing to the changes which occurred in the administrative arrangements in the Province during the period under review, there was bound to be some decline in locally controlled governmental functions, and an increase in those for which the Northern Ireland Office was directly responsible. These changes do not affect the total numbers of employees.

Public sector employment, then, however the term is construed, is of vital importance in the total employment structure. However, a closer look at the statistics shows that, in terms of income maintenance, a shift has taken place which has serious implications. Male employment has dropped since reaching a peak in 1979, and except for the 2,500 rise in male employment in police and prison services, there has been a drop of 5,500 male jobs in departments like Agriculture and Environment, where the loss is particularly serious in districts outside Belfast.

In contrast, the rather large increase in women's jobs (by over

20,000) is heavily concentrated in the category of 'bodies under the aegis of Northern Ireland central government' – that is, education and health services. As we shall see, this means very largely an increase in part-time, low-paid ancillary jobs. We must bear this in mind when we look at the denominational structure of the labour force, because it is precisely in these categories that Catholics are heavily represented.

Even on this cursory first examination, then, the public sector turns out to be not quite such a bastion of solid employment as it may appear at first glance. To be sure, even disregarding the category we have just mentioned (though the female labour force of this category accounts for 44 percent of all public sector jobs), there is still a solid core of positions to be considered, with relatively good prospects, pay, pensions and conditions. The allocation of posts, and the system of promotion, within the public servicc is of the greatest importance, and recruitment and grading practices are also of interest to us. The overall significance of the sector is diminished, however, if we exclude from consideration women's jobs, though even the very small sums earned by them may make a significant contribution to household budgets, especially in the poorer communities.

Overview

In Table 6.4 we present the evolution of the Northern Ireland industrial structure in the last ten years, based on the new (1980) industry divisions. The table essentially corresponds to the category 'civil employment' in the annual statistics, that is, including in this case the self-employed. (For full details see Additional Tables A6.1 and A6.2.)

The trends exhibited in the table are remarkably stable. Agriculture and the small amount of forestry, extractive industry, etc. are stable as a percentage of a now shrinking employed population, so they become relatively more important for the areas outside Belfast. Manufacturing, which will occupy us in detail later, falls drastically, from round about 30 percent to under 20 percent. Male manufacturing employment fell by about a third; that for women halved. Construction, distribution and transport do not fall much as a proportion of the whole job market, and the loss is about equally distributed between men and women. In banking and business services (the new industry divisions replacing the older, more easily identifiable group of banking, insurance and professional services) there is a distinct rise, with men's employment showing a slightly larger increase than women's. In the 'other services' category there occurred a very

Table 6.4 *Industrial structure of Northern Ireland by five new industry division groups, 1974–83*[1]

Industry division groups[2]		1974 Male	1974 Female	1974 Total	1978 Male	1978 Female	1978 Total	1981 Male	1981 Female	1981 Total	1983 Male	1983 Female	1983 Total
I 01–17	(no.)	61,600	6,300	67,900	59,650	8,200	67,850	56,050	8,050	64,100	56,150	8,000	64,150
Agriculture, extractive	(%)[3]	16.6	3.1	11.8	16.5	3.6	11.6	16.6	3.5	11.3	17.6	3.5	11.8
II 21–49	(no.)	111,350	60,850	172,200	95,600	49,550	145,150	82,750	40,800	123,550	69,100	35,600	104,700
Manufacturing	(%)	30.1	29.6	29.9	26.5	22.0	24.7	24.5	17.9	21.8	21.6	15.8	19.2
III 50–79													
Construction,	(no.)	117,350	42,300	159,650	116,050	45,100	161,150	102,200	45,150	147,350	97,400	45,750	143,150
dist., trans., comms	(%)	31.7	20.5	27.7	32.1	20.0	27.5	30.2	19.8	26.0	30.5	20.3	26.2
IV 81–85	(no.)	14,200	9,400	23,600	15,150	11,400	26,550	16,050	12,750	28,800	16,050	12,650	28,700
Banking, business serv.	(%)	3.8	4.6	4.1	4.2	5.1	4.5	4.7	5.6	5.1	5.0	5.6	5.3
V 91–99	(no.)	65,950	86,950	152,900	74,900	111,200	186,100	81,250	120,850	202,100	81,000	123,800	204,800
Other services	(%)	17.8	42.2	26.5	20.7	49.3	31.7	24.0	53.1	35.7	25.3	54.8	37.5
Total	(no.)	370,450	205,800	576,250	361,350	225,450	586,800	338,300	227,600	565,900	319,700	225,800	545,500
	(%)	100	100	100	100	100	100	100	100	100	100	100	100

[1] Figures for 1974 and 1978 by old SICs; for 1981 and 1983 by new industry divisions.

[2] For definition of industry divisions see Appendix B. For further details of these divisions see Northern Ireland Census 1981, *Economic Activity Report*, Table 10.

[3] Percentages may not always add up to 100 due to rounding.

Source: Own calculations based on Northern Ireland *Annual Abstract of Statistics* no. 3, Table 10.2

important rise, from just over 150,000 to over 200,000. Here, the male gain was distinctly smaller than the female increase: men's service jobs rose by 23 percent, women's by 42.5 percent. The *proportionate* gain, however, looking at employment as a whole, was greater for men than for women. Overall the last two groups increase their share from 30.6 percent to 42.8 percent – the men from 21.6 percent to 30.3 percent, and the women from 46.8 percent to 60.4 percent.

The picture remains clear even if we allow for some definitional, accounting year and boundary changes, and there is nothing startling about them. It is however worth pointing out, since this is a time trend table, that by 1978 (the year the employed labour force reached its maximum) a drastic fall in manufacturing employment (by a fifth) had already occurred, and so had most of the rise in the service industries. The decline since 1978 has been due to a further fall in manufacturing (losing another 40,000 jobs, or 28 percent) with only a further small rise in the service industries (10 percent). The latest trends are not encouraging: further quite significant rates of falls in manufacturing of between 5 and 10 percent per annum, and virtual stability in the service sector with a slight tendency to fall.

Table 6.5 shows the 1981 position in somewhat greater detail, emphasizing the importance of part-time employment for women. Over one-third of all employed females worked part-time, and in Group V(d), where 30,000 women worked, two-thirds were part timers: this group includes the great majority of personal services outside the public sector, e.g. catering and domestic services. (For details see Additional Table A3.3).

Local Structure of Industry

In Table 6.6 (and Additional Table A6.4) we examine the evolution of the industrial structure over time, and by area groups (of travel-to-work areas). For this exercise we have to make use of the old classification of industries (by 27 SICs, here grouped in eight sectors which do not exactly correspond to the structure used in previous tables). This is because the data based on the post-1980 classification are not available in sufficient detail. There are confidentiality constraints. As we are dealing with employees, not the self-employed, agriculture (part of the first group) is underrepresented.

We begin with manufacturing. In 1971 this constituted the largest single employment subgroup in all four regions, ranging in importance from 39 percent of male employment in the Belfast group to 29 percent in the Western area. We note that the relative importance of

Table 6.5 *Employees in employment in Northern Ireland, by new industry division groups, males and females, full-time and part-time, 1981*

Industry division groups[1]		Males no.	%	Females no.	%	Female part-time no.	%	Total no.	%
I	01–17	16,628	6.2	1,319	0.9	872	1.2	18,819	3.9
II	21–49	79,512	29.8	36,467	25.5	3,767	5.0	119,746	24.7
III	50–79	78,262	29.3	26,012	18.2	16,442	21.8	120,716	24.9
IV	81–85	13,332	5.0	10,581	7.4	1,987	2.6	25,900	5.3
V(a)	91	34,974	13.1	14,317	10.0	2,052	2.7	51,343	10.5
(b)	92, 95	12,711	4.8	24,762	17.3	12,897	17.1	50,370	10.4
(c)	93	21,171	7.9	18,807	13.2	16,516	22.0	56,494	11.7
(d)	94, 96–9	10,291	3.8	10,531	7.4	20,683	27.5	41,505	8.6
Total		266,881	100.0	142,796	100.0	75,216	100.0	484,893	100.0

[1] For definition of industry divisions see Appendix B.

Source: Own calculations based on Northern Ireland *Annual Abstract of Statistics* no. 3, Table 10.5

this sector was almost as large in the Northern and Southern groups as in Belfast, and dropped to under 30 percent of all male employment only in area group IV.

For women, a similar picture emerges: manufacturing was, in 1971, the largest employment category overall. Here, however, we note that the greatest importance was not in Belfast but in the three other regions, exceeding 40 percent in both the Southern and the Western groups. The overwhelming importance for female employment of textiles, clothing and footwear (SICs XIII and XV) is self-evident. Reading across Table 6.6, we can see the catastrophic decline of employment in manufacturing industry. Remembering that 'employees in employment' (the overall category on which these tables are calculated) hardly changed during these ten years, the percentage variations represent absolute changes, not just relative values. Manufacturing fell by almost 10 percent for males in Belfast, 7 percent in the North, 5 percent in the South and 6 percent in the West. For women the decline was even more drastic. In Belfast their share of employees in manufacturing halved. In the North it fell from 38 percent to 23.2 percent, in the South from 42.8 percent to 22.5 percent, and in the West from 41.8 percent to 23.1 percent. Thus manufacturing, which at its peak in 1974 supported about 170,000 employees, and by 1981 only about 118,000, fell relatively faster for the women than for the men. At the start of the decade some 64,000 women were in manufacturing, by 1981 there were only 40,000, a drop of 37.5 percent. For the men the fall was 28 percent (though, of course, more men than women lost their jobs). We now see how

Table 6.6 *Evolution of industrial structure by pre-1980 standard industrial classifications, in eight groups, by four groups of travel-to-work areas, 1971–81, males, females and totals*

Travel-to-work area groups	1971			1976			1981		
	Male	Female	Total	Male	Female	Total	Male	Female	Total
I Belfast									
Extractive	1.1	0.2	0.8	1.0	0.2	0.7	0.3	0.04	0.2
Manufacturing	39.0	30.9	35.9	34.4	21.0	28.9	30.6	14.6	23.5
Construction	11.2	0.9	7.2	11.2	1.0	7.1	8.3	0.9	5.0
Gas etc., trans. etc.	10.9	2.8	7.8	10.6	2.9	7.4	10.0	2.8	6.8
Distr., ins. etc.	13.4	21.1	16.4	13.5	19.9	16.1	15.1	20.0	17.3
Prof./sci. services	8.3	25.4	14.9	10.1	30.6	18.5	12.7	33.5	21.9
Misc. services	6.9	12.7	9.2	6.5	14.8	9.9	8.1	18.5	12.7
Public admin., defence	9.1	5.9	7.9	12.6	9.5	11.3	14.8	9.7	12.5
II Northern									
Extractive	7.6	1.4	5.3	7.7	1.0	5.1	2.5	0.3	1.5
Manufacturing	37.3	38.0	37.5	35.7	28.9	33.0	30.6	23.2	27.1
Construction	16.4	0.7	10.7	13.8	0.9	8.8	10.1	0.9	5.8
Gas etc., trans. etc.	7.0	2.0	5.2	8.3	1.7	5.7	10.2	1.8	6.3
Distr., ins. etc.	9.0	13.4	10.6	9.7	14.5	11.5	12.9	13.9	13.4
Prof./sci. services	7.8	28.3	15.3	10.6	33.4	19.6	15.0	36.8	25.2
Misc. services	5.0	11.3	7.3	5.8	13.8	8.9	7.4	17.5	12.1
Public admin., defence	9.9	4.9	8.1	8.4	5.8	7.4	11.3	5.6	8.6
III Southern									
Extractive	8.9	1.3	5.9	8.3	0.9	5.2	2.4	0.2	1.4
Manufacturing	35.2	42.8	38.2	32.4	31.0	31.8	30.5	22.5	26.7
Construction	16.7	0.7	10.3	15.2	0.8	9.2	11.0	0.8	6.1
Gas etc., trans. etc.	6.9	1.5	4.8	6.9	1.4	4.6	7.9	1.6	4.9
Distr., ins. etc.	10.8	12.7	11.5	10.2	12.6	11.2	13.0	13.4	13.2
Prof./sci. services	8.5	16.7	15.8	10.9	34.4	20.8	14.5	37.7	25.5
Misc. services	5.5	10.7	7.6	5.5	13.7	8.9	7.4	17.8	12.3
Public admin., defence	7.5	3.6	5.9	10.6	5.2	8.3	13.3	6.0	9.9
IV Western									
Extractive	10.7	0.9	6.8	8.8	0.7	5.6	1.9	0.1	1.1
Manufacturing	29.0	41.8	34.1	25.2	29.0	26.7	23.0	23.1	23.1
Construction	15.2	0.5	9.4	15.9	0.7	9.8	10.9	0.7	6.1
Gas etc., trans. etc.	8.4	1.4	5.6	9.1	1.4	6.0	9.5	1.6	5.8
Distr., ins. etc.	10.7	13.0	11.6	10.0	13.2	11.3	12.9	13.5	13.2
Prof./sci. services	10.2	29.0	17.6	12.1	35.4	21.4	17.1	38.8	27.2
Misc. services	5.7	9.2	7.1	5.6	13.1	8.6	7.7	16.4	11.8
Public admin., defence	10.1	4.2	7.8	13.3	6.5	10.6	16.9	5.8	11.7

Source: Own calculations from unpublished tables provided by the Department of Economic Development, based on Census of Employment for the years shown

unequal the incidence was, and we shall pursue this further when we come to look at unemployment by religion in 1981.

There are other important areas of employment reduction. Construction, distribution and transport (our industry division group III) lost about 10 percent of its employees overall from its peak in 1974, and 12.8 percent of employees in employment (Table 6.4). The subgroups within this were rather unequally affected: construction lost 40 percent of its employees; retailing, hotels and catering actually gained; transport lost 30 percent; and communications gained. We now examine the incidence of this group loss.

Table 6.6 sums up the losses of the area subgroups. Male construction workers had been a relatively smaller part of the workforce in Belfast, but lost a third of their jobs. In the three other areas they were rather more important (15 to 16 percent of all employees) but again lost about a third, which obviously affected the total employment picture rather more.

The public service industries (including gas, electricity, water and transportation, the old SICs XXI and XXII), were more important in Belfast and lost 3,500 jobs, but little in relative importance. They provided less employment in the other areas, and actually gained relatively (that is, a static number of jobs becomes a higher percentage of a shrinking workforce).

Wholesale and retail distribution, which in the old grouping we have ranked with banking, finance and insurance, clearly gains in importance throughout the whole country. The group was most important in Belfast and becomes even more so during the decade, though with little net gain in jobs, men and women benefiting equally (but not, as we shall see, across all subdivisions of this group). In the Northern area, the group was rather underrepresented at the beginning of the decade and assumed average (non-Belfast) importance later on. Again, subdividing this composite group (see Additional Table A6.4) we see that the apparent stability of the group does not exist: distributive trades lose slightly over the whole period but, whereas the men lose overall, the women gain. There is, however, a sharp drop in the distributive employment after 1980.

Banking, insurance and finance are, as we would expect, growth industries, and nationally there is an increase in employment (more definitely identified under the old SIC XXIV than the new industry divisions 81 and 82) which again benefits the women more than the men. Once more Belfast is the main beneficiary but only marginally so, since by 1983 the four business service groups (81–85 in the new classification) only employ 5.3 percent of the working population compared with 4.1 percent in 1974, and the total rise in employment is only 5,000 (or under 1 percent of the employees). This increase clearly does not in any way offset the very large losses in the other groups.

Lastly, we look at the old SICs XXV–XXVII, the various private and public services which are generally held to provide the growth potential to offset the job losses when manufacturing industries decline. In Table 6.6 we have listed these service groups in greater detail, to show changes in employment by gender and locality. At the beginning of the decade, in 1971, the percentages shown represented a total of 166,000 jobs, and this figure had risen to 239,000 by 1981. At first sight this looks as if these increases were sufficient to compensate for losses incurred elsewhere, but this is not how the labour market works.

In Belfast these service jobs had accounted, between them, for nearly a quarter of all male jobs and 44 percent of women's jobs. By the end of the decade the proportion was 35.6 percent of men's jobs in Belfast and 51.7 percent of women's jobs. Outside Belfast their importance had been unequal: over a quarter in the West, rather less in the North, and just over a fifth in the South. For women these sectors mattered as much in the North as in Belfast, and a little less in the South, and less still in the West. By the end of the decade they had reached 50 percent everywhere, so clearly they mattered much more.

If we subdivide these three SICs, their overall increases were about 40 percent in professional and scientific (that is, mostly private) services, with a continued rising trend at the end of the period and into the 1980s. The same sort of increase is found in the 'miscellaneous' group, with a levelling off towards the end, and again there is a similar overall gain in the public administration and defence sector, with the same sort of levelling off overall. In the four groups of travel-to-work areas in Table 6.6 we find the biggest increases in relative importance, in the professional and scientific services, for the women, rising to 38.8 percent of all employment in the West. The men start from lower proportions, and again the biggest relative rise occurs in the West: up from 10.2 percent to 17.1 percent. The 'miscellaneous' group also increased most in the areas where it was previously relatively low in importance, and again the increase was most noticeable for women.

Lastly, the public services, administration and defence group increased everywhere, absolutely and in relative importance. It was, and remained, more important for male employment than for female; it increased by about half for men in the decade, except in the North; and it rose by rather more than that for women, with the highest increase in the Southern group, where it had been lowest overall.

Behind this general picture there lie movements of quite different magnitude. The Belfast labour market comprised, at any point, more than a quarter of a million jobs, and that was always more than half the country's employment. There were as many in the private service group in Belfast as there were women at work altogether in two of the other groups. However, the *impact* of any change in the size or composition of job demand does not depend on the size of the group involved. This works both ways. Because the Belfast labour market is so large, the loss of 20,000 manufacturing jobs may not look so important when in the same period there are increases of 13,000 in service jobs (for men in both cases). However, that presupposes that the other jobs are as easily accessible as the manufacturing industry ones, geographically and from the point of view of qualifications, and that there are no discriminatory barriers. That remains to be seen. Again, in the Western group, the loss of male jobs in manufacturing industry was only about 2,400 while service jobs grew by 4,700, which would make it look as if the men were better off in the rural west and in Derry. In the same group, 2,600 female manufacturing job losses were offset, apparently, by over 8,000 new service jobs.

That these trade-offs do not operate can be guessed from the figures for rises in unemployment, even if common sense did not dictate that such mobility simply cannot exist from any point of view – gender, age, educational attainment or travel-to-work distances, leaving discrimination on religious grounds to one side altogether.

In other words, the rapidly shifting pattern of the labour market's composition in only ten years would have brought severe dislocation even in a perfectly organized world where everybody was free to move between industries and localities. In the circumstances of Northern Ireland, this is clearly utopian. Lastly, it must be pointed out that the defined 'travel-to-work areas' are in one sense misnomers: the whole concept arises from the idea that all jobs which are not purely local and population related are in the geographical (transport) centre of each group. That is clearly not the case, especially when industry declines and service jobs are on the increase. By grouping as many as six travel-to-work areas in one region (as we have done in the South) we are producing something even more artificial. This is why the fuller local figures given at TTWA level and at ESO level (as in Additional Tables A4.2 and A4.3), and in the districts, give a much better idea of the heterogeneity of experience. The analysis of the overview serves merely to produce a first approach to the disaggregation of the figures normally under discussion which relate to the whole of Northern Ireland, men and women, young and old, Catholics and Protestants, car owners and bus users, and unqualified and unskilled as well as those with degrees and diplomas.

The losses we have shown, especially in manufacturing and in construction, are in one sense only gross losses but in another sense also net losses, because there is so little evidence that those who have been made redundant in manufacturing industry have *any* access to the growth sectors. Thus 70,000 fewer employees in manufacturing industry in 1984 compared with 1974, and 15,000 fewer construction jobs (see additional Tables A3.2(a) and (b), new industry divisions 21–50), are not 'made up' by 76,000 more service jobs (new industry divisions 71–99).

Most important, the apparent growth is predominantly in low-paid jobs, and in women's part-time workplaces. Later we will try to analyse further just how little relationship there can be between job losses and job gains. The geography of Northern Ireland is no more inimical to free movement of labour than that of Scotland, Wales or northern England, nor are age and gender differences significantly more pronounced, but religious affiliation plays a very large role.

Unemployment by Industry

In Table 6.7 we try to throw some further light on the problem of overall immobility in the labour market by taking just the thirteen TTWAs and analysing the industries whose decline caused the unemployment to occur. (The industry divisions now revert to the new groupings introduced in the 1980 classification, but the principle is the same.) First we note that Belfast, the largest TTWA with 55 percent of the total employment, has the lowest overall unemployment (17.2 percent), and the smallest TTWA, Strabane, has the highest total (40.3 percent). Secondly, we take note of the large category of NES (not elsewhere specified), that is, mainly those young entrants to the labour market who have never been allocated to an occupational or industrial grouping; at 38,000 in 1981 they constituted 6.8 percent of the total potential working population. They, in turn, are very much more important in the smaller, rural, Catholic TTWAs (plus Derry) than in Belfast and the Protestant North – though this is not a neat pattern because their significance is also relatively small in Fermanagh (Enniskillen), Craigavon, Downpatrick and Dungannon.

Unemployment in industry division group I (agriculture, public services like gas, electricity and water, and extractive industries) is negligible. This is also so in the expanding financial and business service sector, and there are few local differences. The bulk of the unemployment is in manufacturing industry, in construction, distribution and transport, and in the public service sector, despite its apparent expansion. Half the unemployment caused by closures of

Table 6.7 *Unemployed males and females by travel-to-work areas in five new industry division groups, June 1984*

Travel-to-work areas	Estimated total employees	Industry division groups[1]												Percentage of all workers
		I		II		III		IV		V		Not elsewhere specified[2]		
		no.	%	no.	%	no.	%	no.	%	no.	%	no.	%	
Armagh	12,735	97	0.7	440	3.5	808	6.3	35	0.3	548	4.3	1,206	9.5	24.6
Ballymena	47,216	277	0.6	1,999	4.2	3,116	6.6	154	0.3	1,278	2.7	3,018	6.4	20.8
Belfast	306,571	513	0.2	9,959	3.2	16,828	5.5	1,138	0.4	7,791	2.5	16,412	5.4	17.2
Coleraine	25,842	300	1.2	968	3.7	1,793	6.9	67	0.3	839	3.2	1,962	7.6	22.9
Cookstown	6,076	90	1.5	397	6.5	703	11.6	24	0.4	236	3.9	693	11.4	35.3
Craigavon	41,917	254	0.6	1,794	4.3	2,298	5.5	182	0.4	1,620	3.9	1,947	4.6	19.3
Downpatrick	17,739	217	1.2	516	2.9	1,234	7.0	52	0.3	807	4.5	1,159	6.5	22.4
Dungannon	10,856	160	1.5	651	6.0	1,751	16.1	212	2.0	257	2.4	639	5.9	33.9
Enniskillen	16,235	428	2.6	644	4.0	1,304	8.0	51	0.3	746	4.6	1,185	7.3	26.8
Londonderry	41,883	304	0.7	1,806	4.3	3,680	8.8	136	0.3	1,366	3.3	4,837	11.5	28.9
Newry	18,682	157	0.8	786	4.2	2,599	13.9	48	0.3	638	3.4	1,970	10.5	33.1
Omagh	12,866	138	1.0	261	2.0	863	6.7	28	0.2	454	3.5	1,299	10.1	23.5
Strabane	9,251	220	2.4	514	5.6	1,156	12.5	24	0.3	232	2.5	1,584	17.1	40.3
Northern Ireland	567,869	3,155	0.6	20,735	3.7	38,133	6.8	2,151	0.4	16,812	3.0	37,911	6.8	21.3

[1] For industry divisions see Appendix B. For groups see Table 6.4.

[2] These are the unemployed who cannot be allocated to any industrial group; they are mostly young workers who have never worked in any industry.

Source: Own calculations based on Northern Ireland *Annual Abstract of Statistics* no. 3, Table 10.10, and Department of Economic Development Press Notices, April 1984

manufacturing industry is in Belfast, and another 18 percent in Craigavon and Derry, with Ballymena as the only large Protestant centre of unemployment. Belfast has also half the unemployment in most other groups, except construction, transport and distribution where its share is rather smaller than expected, with only 44 percent of the total. The figures in this table highlight the problems of mobility. Even where unemployment is low overall, as in the services group, Belfast has only 2.5 percent of its workforce displaced from that sector. Those presumably looking for work (7,700) as the result of this displacement alone are almost as numerous as those from all other areas together. It seems very unlikely therefore that such workers could find jobs elsewhere in Northern Ireland, even without allowing for the fact that they are unlikely to have the skills required to find positions in the growth areas of the service sector.

Summary

During the period under review, the structure of labour demand has changed considerably. Within the self-employed sector, there has been a movement away from the agricultural sector into some forms of specialist and craft production of goods. Retailing, too, has lost in self-employment, though not as much as in agriculture. The future outlook is poor: at most, one would expect self-employment to hold its share of the total market demand.

The public sector is much more important. It was the most buoyant part of the total labour market in the 1970s, as against decline in most other sectors. This growth was partly due to the rise in police and other security services, and partly due to increases in health, education and social services. Some of these were related to population, some to a genuine increase in service levels. Recently this rise has levelled off, and a decline has in fact set in. To some extent, employment in this sector must fluctuate with levels of perceived security: if the Province experiences less strife, male employment in particular will drop. Female employment is more at risk as the result of changes in decisions about the level of personal public services. Female public sector employment has been heavily concentrated in low-paid and part-time jobs, especially in health and education, and the total effect on employment and incomes is therefore more serious if male employment declines in this sector, as it has done.

Agriculture has remained stable, or declined only slightly, from its already low levels in the mid 1970s. Manufacturing has suffered the greatest losses: it accounted for 30 percent of the workforce in 1976, and only 20 percent in the most recent year.

The banking and financial sector gained considerably, again more for women than for men. The 'other' services (that is, private sector personal services) showed the biggest increase, by a third, but again most of that accrued to unskilled women workers, a high proportion in part-time jobs.

The local incidence of employment changes reflects these differences. The core areas most heavily dependent on manufacturing, plus the Londonderry travel-to-work area, suffered the greatest loss in manufacturing; but whereas Greater Belfast also had considerable growth in the service sectors, this did not accrue to the peripheral areas. The lower-paid general service sectors grew relatively at the same rate throughout the Province, but the best-paid jobs (in banking, insurance, and finance) grew significantly only in Greater Belfast.

Greater Belfast, in fact, accounts for roughly half the jobs in the Province, and this proportion does not change over time. But unemployment is lowest in the capital area, and the proportion of well-paid jobs, and openings for people with professional and high technical qualifications, is much greater there than elsewhere. Nevertheless, in Belfast the loss of manufacturing jobs was greater than the increase in service jobs; in other areas the proportions are reversed. Since people losing manufacturing employment cannot easily be transferred to service occupations, especially of the more highly skilled variety, the unemployment problem may be more intractable in Belfast than elsewhere. The loss, in ten years, of 74,000 manufacturing and 15,000 construction jobs is in no sense made up by an increase of 76,000 jobs in the service sector: not by age, by gender, by area or by religion. In other words, the local as well as the national mismatch between labour availability and labour demand has increased more sharply than overall employment figures suggest. The percentage loss to household incomes from these movements is much larger than the percentage rise in unemployment.

7

Industrial Employment Structure, Employment Status and Religion

Industrial Employment and Religious Structure

We must now take a more detailed look at employment by industry, area and religious affiliation. As the 1981 Census (our sole source of information on religion) analysed employees in the new industrial divisions, the figures here cannot be compared in detail with some of the earlier tables of the evolution of employment and unemployment, except for manufacturing industry (mainly the new division groups II, III and IV).

Table 7.1(a) shows, for males, the percentage distribution of employees between industries, for all denominations, and for Roman Catholics. (It has to be remembered that this analysis includes the 'not stateds' in the 'all' column, who comprise varying proportions of Catholics. Not all industrial divisions are shown.) The object here is to show that the industrial structure of the Catholic community not only is radically different from the population as a whole, but also varies locally. Looking at the last column, we immediately notice the *over*-representation of Catholic males in construction (division 5), in catering (66), in wholesale distribution (61) and in education (93). On the other hand, they are *under*-represented in the very important division 3 (metal etc. manufacture, engineering), especially the sub-group 36 (other transport equipment, that is, shipbuilding and aircraft), in the whole public administration group (91) and to a lesser extent in banking, finance etc. (8).

These differences are accentuated at the local level. In Belfast the underrepresentation in metal manufacture, ships and aircraft is even more striking, as we would expect; and there is the same imbalance in public administration. There is the same overrepresentation in construction, in hotels and catering, but also a more pronounced excess in transport and communications. There is also relatively high employment in education and medical services.

In other areas the pattern is repeated, with small variations. Only in the West is Catholic representation in industry a little larger than proportionate, but there is more pronounced underrepresentation in the service sector, most clearly so in the public services.

Table 7.1(a) *Roman Catholic males and all males in main industry divisions as percentage of all males in employment, by area group, 1981*[1]

Industry	I Belfast		II Northern		III Southern		IV Western		Northern Ireland	
divisions[2]	All	RC only	All	RC only	All	RC only	All	RC only	All	RC only
1	1.95	1.4	12.0	11.3	13.6	11.2	15.2	11.7	8.0	8.5
3	15.4	8.6	5.6	5.2	4.1	3.6	2.4	3.2	9.5	5.2
32	3.8	2.0	1.3	1.5	1.0	1.0	1.3	1.8	2.4	1.5
36	7.7	1.8	0.5	0.4	0.6	0.3	0.2	0.4	3.9	0.8
4	10.5	8.9	12.5	10.8	14.9	13.4	10.8	12.1	11.8	11.4
41, 42	4.1	3.5	5.5	3.6	5.1	4.8	4.3	4.5	4.6	4.2
43	1.6	1.0	1.8	0.8	3.1	2.1	2.4	3.1	2.1	1.9
48	1.6	1.8	2.7	3.2	2.1	2.2	0.7	0.9	1.7	1.9
5	9.8	14.1	12.8	17.8	14.4	18.9	11.9	14.3	11.6	16.2
6	16.0	16.0	13.8	14.7	14.5	15.1	13.0	14.5	14.9	15.1
61	5.2	3.9	3.3	2.7	3.4	3.3	2.3	2.2	4.1	6.3
64, 65	7.2	7.2	6.7	6.8	7.2	7.3	6.9	7.5	7.1	7.3
66	1.4	2.9	1.4	2.6	1.4	2.1	1.6	2.4	1.4	7.5
7	6.8	10.7	6.2	6.3	4.6	4.5	4.1	4.6	5.9	6.6
72	2.3	3.5	2.6	2.6	2.2	2.2	1.6	1.8	2.2	2.5
79	2.6	4.6	1.6	1.7	1.8	1.5	1.9	2.2	2.2	2.6
8	6.6	7.0	4.0	3.5	4.0	3.6	4.0	4.1	5.2	4.6
83	3.9	4.3	2.5	2.2	2.4	2.3	2.5	2.9	3.1	3.0
9	26.3	27.7	21.3	20.2	21.6	21.0	26.6	22.4	24.5	23.2
91	13.9	10.4	11.1	8.1	11.0	7.3	14.8	8.1	12.9	8.5
93	5.5	8.2	5.0	6.1	4.7	6.5	5.6	7.6	5.3	7.2
95	2.7	3.8	2.3	2.6	2.9	3.5	3.1	2.8	2.7	3.3
96	1.4	1.5	1.2	1.3	1.3	1.4	1.2	1.5	1.3	1.4
Total in employment	153,595	20,720	53,200	9,875	70,021	22,820	45,299	18,333	322,115	71,748

[1] For definition of area groups see Appendix A.

[2] For definition of industry divisions see Appendix B. Single figures denote major industry divisions, double figures subdivisions.

Source: Own calculations based on Northern Ireland Census 1981, *Religion Report*, Table 9 (unpublished information)

At first sight these imbalances may not seem very dramatic. It is only when we read these figures in conjunction with what information we have about employment status (for example, in education: school teachers or caretakers?) that we shall see the true implications of the analysis.

When we turn to the female industrial structure (Table 7.1(b)) we see a different pattern. Industrial employment is now more evenly distributed. Serious underrepresentation occurs, nationally, in retail distribution, and in footwear and clothing. This is apparently compensated for by overrepresentation in the service sector, where we distinguish, clearly, between public administration where there is some underrepresentation, and education and health services where the proportions are more favourable to Catholics. At the local level there is serious underrepresentation in Belfast in all industries except clothing and footwear, and in distribution. On the other hand, we again have the pronounced higher percentages in the services sector, with apparent parity even in the public administration division.

Outside Belfast, the imbalances are similar. In the West, the services sector employs fewer Catholic women than we would expect from the national picture, but there is relatively more employment in industry. The growth sector of banking, finance etc. (division 8), which has only a small imbalance by religious group nationally, has relatively worse rates for Catholic women both in the predominantly Protestant Northern group, and in the West, though the numbers involved are very small.

Once again, the most important question will be: what is the status of the employees in these industrial divisions? Additionally, for women we need to ask: how many of them are part-timers (and therefore most likely to have low status and low pay)?

In Table 7.2 we attempt a fresh approach, for women, to the question of employment status. We single out six industry divisions which between them employ over 60 percent of all females in employment. In each case, married women constitute more than half of those in the industry, the figure rising to 63 percent in hotels and catering, 71 percent in education, and 74.5 percent in 'other services' to the public. From this table we derive the view that these must be the older women, and also that the majority of them must be part-time workers. However, these are Census figures, and we had some suspicion that this table might understate the true position. We therefore also obtained (Table 7.3) a breakdown of employment known to the Department of Economic Development, which provides a summary of persons working (at the time of the Census of Employment in 1981) in certain specified categories. These people are included in the total employment statistics we have been using for

Table 7.1(b) *Roman Catholic females and all females in main industry divisions as percentage of all females in employment, by area group, 1981*[1]

Industry divisions[2]	I Belfast		II Northern		III Southern		IV Western		Northern Ireland	
	All	RC only	All	RC only	All	RC only	All	RC only	All	RC only
3	4.2	1.7	3.7	2.7	2.2	1.8	2.3	3.3	3.5	2.1
34	1.6	0.5	2.5	1.8	1.2	1.2	1.8	2.3	1.7	1.3
4	11.2	8.2	17.1	15.6	18.8	15.5	19.5	22.5	14.7	14.8
45	3.4	3.5	6.7	8.3	8.2	8.0	12.7	15.2	6.1	8.3
6	19.4	15.7	17.1	14.8	14.9	12.3	14.8	12.9	17.5	13.9
61	2.9	1.8	1.7	1.0	1.5	1.1	1.2	0.8	2.2	1.2
64, 65	12.7	10.1	11.3	8.8	10.1	8.3	10.1	8.9	11.7	9.1
66	3.2	3.6	3.8	4.8	3.0	2.8	3.3	2.9	3.3	3.3
7	2.2	2.6	2.0	1.5	1.6	1.4	1.7	1.4	2.0	1.8
79	1.3	2.0	1.0	0.8	1.1	0.8	1.4	1.1	1.2	1.3
8	7.4	7.1	4.9	3.6	4.8	4.5	4.5	3.8	6.1	5.1
81	2.3	2.4	1.9	1.5	1.9	1.7	1.7	1.4	2.1	1.8
83	3.8	4.0	2.4	1.7	2.4	2.5	2.3	2.0	3.1	2.8
9	51.1	57.3	49.5	57.4	51.7	58.8	51.2	50.7	51.0	57.3
91	12.6	12.6	8.9	8.8	9.1	8.2	8.7	6.9	10.8	9.6
93	12.8	15.6	15.7	19.8	14.2	18.7	17.3	18.3	14.3	17.7
95	14.3	20.3	15.2	19.4	16.4	19.5	15.0	14.6	15.0	18.6
96	6.7	6.5	5.5	4.7	6.9	7.4	6.4	6.7	6.5	6.6
98	1.8	2.3	1.7	2.2	1.7	2.0	1.4	1.7	1.7	2.1
Total in employment	104,710	16,244	30,635	6,200	41,326	14,708	25,508	11,746	202,179	48,898

[1] For definition of area groups see Appendix A.

[2] For definition of industry divisions see Appendix B. Single figures denote major industry divisions, double figures subdivisions. For any column, only main divisions add up to approximately 100 percent.

Source: Own calculations based on Northern Ireland Census 1981, *Religion Report*, Table 9 (unpublished information)

Table 7.2 *Females in employment, distinguishing married women, in Northern Ireland, by selected industry divisions, 1981*

Industry divisions	Females employed in division(s)	As % of total females employed	% married women employed in division(s)
64, 65 Retail distribution	23,555	11.7	57.9
66 Hotels, catering	6,614	3.3	63.0
91 Public administration, defence	21,877	10.8	54.4
93 Education	28,954	14.3	71.0
95 Health, veterinary services	30,269	15.0	59.6
96 Other services to public	13,212	6.5	74.5

Source: Northern Ireland Census 1981, *Economic Activity Report*, Table 9

Table 7.3 *Male and female employment in subdivisions of industry division 96 'other services to the public', 1981*

Industry subdivisions	Male		Female		
	Full-time	Part-time	Full-time	Part-time	Total
9611 Social welfare, charitable, community services	1,459	158	5,359	17,430	24,406
9631 Trade union, business/professional associations	327	40	214	90	671
9660 Religious organizations, similar associations	435	265	172	309	1,181
9690 Tourist offices, other community services	659	976	661	933	3,229
96 Other services to the public	2,880	1,439	6,406	18,762	29,487

Source: Own calculations based on Northern Ireland Census 1981, economic activity tables (unpublished information)

our overall analysis, but they cannot be broken down by religion. From this table we identify a very large group of women (23,000, most of them part-timers) who were employed in a subgroup 9611 (social welfare, charitable and community services), an omnibus category covering a very large range of low-paid jobs. The discrepancy is seen when comparing the Census totals for division 96 with the DED totals: for men the Census gives something over 4,000, which is not totally inconsistent with the September 1981 figure of 2,880 full-timers and 1,439 part-timers. But the Census figure for females (13,212 total, of whom 9,841 were married) cannot be reconciled with the figure of over 25,000 full-time and part-time women shown in the Census of Employment. One suspects, therefore, from this limited evidence, where a direct comparison can be made, that in certain occupations there is a fairly large group of employees who do not regard themselves as being employed, or perhaps have more than one part-time job and are therefore double-counted in the employment census. (This problem has already been referred to in Chapter 5.) We should therefore be cautious in accepting figures in the various service categories at their face value: the Census of Employment may well overstate the number of real jobs there are, and understate unemployment. These figures by themselves certainly provide only limited guidance for the economic position of women, in particular, let alone household incomes.

Employment Status and Religion

National Overview

In Chapter 5 we looked briefly at the 1981 structure of the labour force in terms of occupational status (see Tables 5.1 and 5.2). We must now return to this matter in greater detail, because we have seen the limitations of the industrial analysis. In many cases it appears that the distribution of industrial employment by religion does not vary very greatly, apart from some obvious and long-recognized disparities, like the absence of Catholics from large sections of the engineering industry, from the security services, and from important branches of public administration. In some industry divisions, notably education and health, there seems to be a disproportionately large amount of Catholic employment, and this also applies to certain private sector industries.

It is therefore most important to see what status these employees have within the industries concerned, as far as that can be done from the available population Census analysis. The Census of Employment does not distinguish religious affiliation; nor do the monthly Department of Economic Development unemployment statistics.

The 1981 Census is therefore our sole source of information, and even that is limited. It does show, however, the proportion of people in the higher positions (managerial, professional and supervisory status in particular) and, even within the main employment divisions, people in more highly skilled occupations are often distinguished from the less skilled. A great deal depends, of course, on accurate assessment (or self-assessment) and there are well-known pitfalls in this analysis: nevertheless, one would hope that when there are half a million workers involved, something positive can be said.

Perhaps one should repeat the usual caveats before presenting some of the figures. The Census response was limited. Within the returns which were received, there was a large number of people whose religion was not stated, and we have no means of allocating this last category by any mathematical manipulation to one religious group or another overall, let alone within the many subcategories which appear in the occupational status tables.

Nevertheless, the data available are not quite useless because, in order for the proportions (at least for the larger categories) to be seriously distorted, one would have to assume that the non-enumerated people in particular differed seriously from those analysed, from the point of view of policy guidance. For instance, a higher proportion of those whom the Census failed to catch altogether may have been unemployed: this would mean that the conclusions in this section are understatements of the true differential. It is unlikely that the 80,000 Northern Ireland residents not included in the Census included a higher than average proportion of professional, managerial or supervisory grades. On the other hand, it is quite possible that the 'not stated' categories included more people, of all religions, in the higher echelons, and for this there is some evidence, but it tells us nothing about possible bias. And again we note that the unemployed proportions for the 'not stateds' are similar to those for Catholics.

So, on all these counts, we have to make the assumption that the available Census printout does not contain any systematic bias. However, we have to be more than usually careful not to make too much of very small absolute or percentage differences: there is, indeed, every likelihood that 4 percent of one category for Catholics is significantly different from 8 percent for 'others', but not that 28 percent for Catholics differs markedly from 32 percent for 'others'. We bear this in mind as we proceed with our analysis.

In Table 7.4 we measure the employment and occupational structure. We compare the percentages of each denominational group who are in different categories of the employment status tables. This measure is in our view the best overall way of showing how the Catholics differed from the Protestants and the 'not stateds'.

One of the by-products of these tabulations is once again to raise the question of whether the 'not stated' category can be allocated in any meaningful way. It is evident from Table 7.4 that the 'not stated' group has overall unemployment rates very similar to those of Catholics, and quite different from those for Protestants. But we should be very chary of concluding from this piece of evidence that the great majority of 'not stated' Census respondents were Catholics, because as we have seen in so many tables, and again recognize from many subcategories in Table 7.4, they may in other respects be quite unlike Catholics, and there are differences in area groups. Nevertheless, it is

Table 7.4 *Economic position by employment status, religion and sex, Northern Ireland only, 1981 (percentages)*

Employment status	Roman Catholics Male	Roman Catholics Female	All other denominations Male	All other denominations Female	Not stated Male	Not stated Female
Total inactive	23.6	59.6	22.8	58.3	19.1	56.7
Retired	11.1	11.3	14.3	12.7	9.5	8.6
Students	7.4	7.4	5.2	4.8	6.2	6.2
Others	5.1	40.9	3.3	40.8	3.4	41.9
Total active	76.3	40.4	77.2	41.7	80.8	43.5
Out of employment	23.1	6.9	9.6	4.0	19.3	6.9
In employment	53.3	33.5	67.6	37.7	61.4	36.5
Self-employed	10.0	0.8	10.6	1.0	9.4	0
Employees	43.3	32.7	57.0	36.7	52.1	35.7
Managers (large firms)	1.5	0	3.7	0	2.5	0
	2.9	**1.2**	**5.4**	**1.8**	**4.0**	**1.6**
Managers (small firms)	1.3	0	2.4	0	1.8	0
	2.5	**0.9**	**3.6**	**1.6**	**3.0**	**1.2**
All managers	**5.4**	**2.1**	**9.0**	**3.4**	**7.0**	**2.8**
Manual foremen	1.7	0	2.7	0	2.1	1.7
	3.2	**0.6**	**4.0**	**0.7**	**3.3**	**0.7**
Non-manual foremen	1.2	1.5	1.8	1.8	1.8	1.7
	2.2	**4.6**	**2.7**	**4.9**	**2.9**	**4.7**
Apprentices	2.1	0	1.8	0	1.8	0
Professional employees	1.4	0	2.7	0	2.2	0
	2.7	**0.8**	**4.0**	**0.9**	**3.6**	**1.0**
Other employees	34.0	29.8	41.9	32.5	39.9	31.9
Total population 16+[1]	134,557	146,160	287,608	317,375	90,838	92,132

[1] Percentages in each row are calculated on total population 16+ in that denomination and gender. Bold numbers represent percentages based on the total of *all* those in employment.

Source: Own calculations based on Northern Ireland Census 1981, *Religion Report*, Table 8 (unpublished information)

a remarkable feature of *all* areas that the 'not stated' group's unemployment is more nearly like that of the Catholics than of the Protestants. It is equally clear that the degree of congruence varies from area to area, and that therefore we could not use the same breakdown locally as that which we might assume nationally.

In Additional Table A5.3 we have calculated the occupational structure by sex and denomination and by districts and area groups. In this more detailed table we have also shown the percentages in two ways: not only what proportion of men and women in the two main religious groups were found in the different status groups, but also what proportion of all employees in the main status groups belonged to the Catholic, 'all other', and 'not stated' groups. In this chapter we reproduce the analysis only for the Belfast area group (Table 7.5), and the commentary for this area, to show general trends for the most important concentration of workers.

We start by looking at activity rates again (Table 7.4): 23.6 percent of Catholic males over 16 were economically inactive, 22.8 percent of 'others', and only 19.1 percent of the 'not stateds'. For females, 59.6 percent of the over 16s were inactive among Catholics, 58 percent of the 'others', and 56.6 percent of the 'not stateds'. Clearly there is not enough in these figures to discern a pattern.

If we now turn to the economically inactive subcategory of 'retired', we find 11.1 percent of Catholic males in that position, 14.3 percent of the 'others' and 9.5 percent of the 'not stateds'. This is significant – but in what way? Is it due to the younger Catholic age structure (even omitting the under 16s as we do in this calculation)? It cannot be due to the proportion of early retirers, because that, as we have seen in Tables 5.2 and 4.3, is high among Catholics. And the 'not stateds' cannot be predominantly Catholics if one used that criterion. Among females, differences are again too small, and the whole concept too vague, to make useful comparisons.

The 'student' category shows the expected dominance among Catholics: 7.4 percent of the over 16s were still in full-time education, compared with 5.2 percent of 'others' and 6.2 percent of 'not stateds'. (We do not subdivide the 'others', but in this case the particularly low rate among adherents of the Church of Ireland (4.9 percent) may be significant; we do not know the answer.) Of the Catholic women, 7.4 percent were students, 4.7 percent of 'others' (4.5 percent of Church of Ireland women), and 6.1 percent of the 'not stateds'. Again these facts need to be interpreted with caution. The age structure of the religion groups differs: there are relatively more Catholics in the age groups 15–24 than among the 'others' (see Figure 23) and the age-specific educational participation rates are in fact rather similar. We shall return to this problem later.

There is a residual category of the economically inactive: 'others'. This means largely the permanently sick among the men; for the women it mostly means just women who did not work, and that is not a category we need analyse further. But for the men, 5.1 percent of Catholics fell into that category, but only 3.3 percent of the 'others' and 3.4 percent of the 'not stateds'. Though absolute numbers are small, this may be significant.

When we turn to unemployment, the differences become more pronounced: 23.1 percent of Catholic men were unemployed, 9.6 percent of 'others' and 19.4 percent of the 'not stateds'. Unemployment among Presbyterians was 8.3 percent. Of the Catholic women, 6.9 percent returned themselves as unemployed, 4.0 percent of the 'others' and 6.9 percent of 'not stateds'. If we compare Census unemployment (76,000 males and 29,000 females) with the claimant-based unemployment, the figures are entirely credible. Allowing for those who did not find their way into the Census at all (and who may well have included a high percentage of unemployed persons), the Census figures are still if anything on the high side, especially for women.

So clearly here we have a real difference; it is one that underlies many of the conclusions of this report, and indeed of other investigations. We shall see what all this means at the more local level.

The self-employed, once again, do not show marked differences at the national level: 10 percent of Catholic men were in that category, only marginally fewer than for the 'other' categories, and more than for the 'not stateds'. We do not here print the subcategories of self-employed with or without employees: 25.6 percent of self-employed Catholics had employees and 29.5 percent of Presbyterians, and this would hardly seem to matter.

Turning now to the managerial, foreman and professional groups, the differences are again fairly evident. (Note: foremen shall be taken to include forewomen; the other categorizations are not gender specific.) Only 1.5 percent of Catholic men were managers of large concerns, and 1.3 percent of small ones; this is 2.9 percent (allowing for rounding) of all Catholic men. This, of course, was already apparent when we looked at socio-economic structure overall (see Chapter 5). Of 'other' men, 3.6 percent were managers of large firms, 2.4 percent managers of small firms; this is 6.1 percent all told, or twice the proportion of Catholics. Of the 'not stateds', 4.3 percent were managers of some sort: this is a little nearer Catholics than 'others'.

It might be more illuminating to count managers only as a percentage of all employed persons, to get a better overview, thus discounting early retirement and unemployment. The figures then show some shrinking of the differential: 5.4 percent of Catholics were managers of some sort, 9 percent of 'others', and 7 percent of 'not stateds'. (See bold figures in Table 7.4, and Table 7.6.)

Since only just over 1 percent of all adult women were managers there seems no point in pursuing the analysis. Probably the denominator should again be 'women economically active and in employment': then the proportion of managers would be nearly 3 percent overall, 2.1 percent for Catholics, 3.5 percent for 'others' and 2.8 percent for 'not stateds'. We should read that part of the Census analysis as meaning: women do not get a proportionate share of management jobs, and Catholic women are particularly disadvantaged. But since the first cause far outweighs the second, we should not pursue the analysis too far.

When we turn to foremen (forewomen, supervisors, leading hands etc.) in manual occupations, Catholic men have 1.7 percent of their over 16s in such jobs, and Catholic women a negligible number. This compares with 2.7 percent for 'others' and 2.1 percent for the 'not stateds'. Using again only the men in employment as the denominator, which is probably more relevant for status questions, the Catholic male proportion rises to 3.2 percent, that for 'others' to 4.0 percent and that for 'not stateds' to 3.3 percent, so the differential somewhat shrinks when one allows for economic inactivity and unemployment rates.

The non-manual foremen, supervisors etc. present a similar picture. For Catholic men they form 1.2 percent of the total adult population and 2.2 percent of those in employment. For 'others' the proportions are 1.8 percent of the over 16 population and 2.7 percent of those in employment. For the 'not stateds' the proportions are 1.8 percent of all over 16s and 2.9 percent of the employed. So the differential once again persists in slightly attenuated form.

Women are more heavily represented among the non-manual supervisory grades (mostly in offices and public administration). They formed 1.5 percent of the adult population and 4.6 percent of those in employment. For the 'not stateds' it was 1.7 percent of the total group and 4.7 percent of the employed. For 'others' they formed 1.8 percent of the over 16s and 4.9 percent of those in employment. So from this one would say that within this not insignificant type of employee, Catholics were not seriously underrepresented.

The category of 'apprentices' looks small in this table when compared with the whole working population, and differences do not have much significance at this level. There are no striking differences by religion, or by sex. Inspection of the more detailed figures (see also Table 7.5) shows that there were some women apprentices, probably mostly in trades like hairdressing, so that the job prospects of young people in this category cannot be gauged from the totals involved.

In fact, apprentices would be better measured as a percentage of all employees in the relevant age groups, rather than the whole working population. On this calculation, nearly 15 percent of boys in the employed age group 16–24 were apprenticed, and the proportion was as high as 40 percent for the 17 year olds and 34 percent of the 18 year olds. For girls the percentages are lower: 10 percent of those aged 16, 17 or 18, and in employment, were apprenticed. These figures are high compared with the position in Great Britain. Why this should be so cannot be explained here, but it is clearly an important subject for a future investigation.

We can in fact take this analysis a little further. We may add up students and apprentices for the denominational groups and work out their proportions to the 16–24 age group, so as to get a sense of what percentage had any kind of reasonable prospects in the labour market. The result is that we have 10,028 Catholic students plus 3,105 apprentices, which makes 13,133 boys against an age group total of 37,000, which equals 35.5 percent; again, by British standards, this is a respectable proportion. For the 'others', this calculation yields almost exactly 20,000 students plus apprentices against an age group of just under 60,000, so we have a slightly lower percentage of 'others' with fair prospects compared with Catholics. It may again be objected that the Catholics are more seriously underenumerated, and we would agree: but we have no evidence that among the non-enumerated and non-respondents the students and apprentices are less well represented. We shall look at this point again at the local level, but on the evidence this is a field where one would have thought that the chances of the Catholics in the labour market should be about equal. However, for all we know this may have been true in the past, and yet current unemployment and employment status figures are discouraging.

Lastly, we touch on the employment status category listed as 'professional'. This group has clearly the best prospects: it includes almost all the technical cadres, the future managers and probably a high proportion of future self-employed professionals. Nationally they account for 3.6 percent of those in employment. For Catholics the proportion was only 2.7 percent, and it was nearly 4 percent for the 'others' and 3.6 percent for the 'not stateds'. For women the figures are again too small for analysis or for significant differences to be noted: the proportion was under 1 percent nationally, and very similar for Catholics and 'others'.

These figures should be read in conjunction with what we have just noted about students and apprentices, and with earlier remarks about managers. Clearly, the education and training position in 1981 was not too unequal for Catholics (we shall discuss this again in Chapter 9); but this is not reflected in the jobs hierarchy as it existed in 1981. We do not know, of course, whether twenty or thirty years ago the position of the Catholics in the younger age groups was equally promising.

The Belfast Area

Though Belfast is not in any way representative of the whole Northern Ireland labour market, we adduce the occupational analysis here because it illustrates some of the basic weaknesses of the Catholic position across the board. All industries and occupations are present in sufficient numbers to allow a detailed analysis.

In Table 7.5 each column shows the number (and percentage below) of persons in each category by religion. This analysis refers to the population over 16, that is those who could be economically active, and those who have retired from the workforce. Thus the table shows that 8.4 percent of all Catholic men over 16 were counted as students, and 21.3 percent were out of employment. This enables us to show differences in status. (The columns add up to more than 100 percent since people can belong to more than one status group.) Table 7.5 also shows the percentages of people in each religious group related to the total number of persons in the relevant employment status group; thus 68.7 percent of all persons in employment belonged to the 'all others' denominational group.

We have here the expected Catholic excess share of unemployment, but with the 'not stateds' exactly halfway between Catholics and 'others' (Table 7.5). The share of managers and foremen is as small as we would expect, with only the non-manual women coming anywhere near their expected proportion. Professional grades are again badly represented.

The alternative approach (Table 7.6) is to see how the structure looks when we exclude the inactive and the unemployed. The result is, throughout, that the share of the higher grades of employees goes up somewhat, but not enough to bridge the gap. Thus if we take all male Catholic managers (1,588 in Greater Belfast) and relate them to those in employment, they form 7.6 percent instead of the 4.3 percent of the over 16s; for 'others' they form 10.8 percent instead of 7.3 percent; for 'not stateds' they form 9.2 percent instead of 6.2 percent. If we took as the denominator the numbers of employed persons excluding self-employment, we would get 8.5 percent for Catholics, 12 percent for 'others' and 10.2 percent for the 'not stateds'.

It is a moot point whether we choose one or the other denominator. The percentage of persons in these grades against all adults gives

Table 7.5 *Economic position by employment status, religion and sex for the Belfast area, by numbers and as percentages of the population over 16, 1981*

		Roman Catholics		All others		Not stated	
		Male	Female	Male	Female	Male	Female
Total inactive	(no.)	7,975	23,497	35,187	100,839	7,187	22,297
	(%)	21.9	54.8	22.7	56.4	17.6	52.0
Retired	(no.)	3,338	4,965	22,525	22,953	3,434	3,476
	(%)	9.1	11.6	14.5	12.8	8.4	8.1
Students	(no.)	3,075	3,019	8,120	7,797	2,653	2,490
	(%)	8.4	7.0	5.2	4.4	6.5	5.8
Others	(no.)	1,562	15,513	4,542	70,089	1,100	16,331
	(%)	4.3	36.1	2.9	39.2	2.7	38.1
Total active	(no.)	28,506	19,418	119,782	77,944	33,598	20,581
	(%)	78.1	45.2	77.3	43.6	82.4	48.0
Out of employment	(no.)	7,786	3,174	14,348	7,265	6,157	2,794
	(%)	21.3	7.4	9.3	4.1	15.1	6.5
In employment	(no.)	20,720	16,244	105,434	70,679	27,441	17,787
	(%)	56.8	37.9	68.0	39.5	67.3	41.5
Self-employed	(no.)	2,092	260	10,306	1,568	2,655	334
	(%)	5.7	0.6	6.7	0.9	6.5	0.8
Employees	(no.)	18,628	15,984	95,128	69,111	24,786	17,453
	(%)	51.1	37.2	61.4	38.7	60.8	40.7

Table 7.5 *continued*

		Roman Catholics		All others		Not stated	
		Male	Female	Male	Female	Male	Female
Managers (large firms)	(no.)	877	243	6,837	1,327	1,466	338
	(%)	2.4	0.6	4.4	0.7	3.6	0.8
Managers (small firms)	(no.)	711	209	4,542	1,230	1,058	269
	(%)	1.9	0.5	2.9	0.7	2.6	0.6
Manual foremen	(no.)	600	69	4,517	457	949	86
	(%)	1.6	0.2	2.9	0.3	2.3	0.2
Non-manual foremen	(no.)	644	868	3,362	3,682	1,055	971
	(%)	1.8	2.0	2.2	2.1	2.6	2.3
Apprentices	(no.)	807	455	3,144	987	798	225
	(%)	2.2	1.1	2.0	0.6	2.0	0.5
Professional employees	(no.)	789	188	4,918	768	1,339	241
	(%)	2.2	0.4	3.2	0.4	3.3	0.6
Other employees	(no.)	14,200	13,952	67,808	60,660	18,121	15,323
	(%)	38.9	32.5	43.8	33.9	44.4	35.7
Total population 16+[1]	(no.)	36,481	42,915	154,969	178,783	40,785	42,878
	(%)	15.7	16.2	66.7	67.6	17.6	16.2

[1] Total population: male 232,235; female 264,576.

Table 7.6 *Economic position by employment status, religion and sex for the Belfast area, as percentages of all those in employment, 1981*

	Roman Catholics		All others		Not stated	
	Male	Female	Male	Female	Male	Female
In employment	13.5	15.5	68.7	67.5	17.9	17.0
Managers (large firms)	4.2		6.5		5.3	
Managers (small firms)	3.4	2.8	4.3	3.6	3.9	3.4
Manual foremen	2.9		4.3		3.5	
Non-manual foremen	3.1	5.3	3.2	5.2	3.8	5.5
Professional employees	3.8		4.7		4.9	

Source: Own calculations based on Northern Ireland Census 1981, *Religion Report*, Table 8 (unpublished information)

a better overview of the total income and status structure for the population. The smaller the denominators are, the better they serve as indicators if we are trying to measure the chances of advancement within the work situation.

In Table 7.6 we give the alternative denominators for some of the grades of employees, so the reader may judge the impact of different methods of calculation. Table 7.5 shows the overall socio-economic structure, whereas the alternatives refer only to the composition of the labour force at work.

The same analysis applies to the foremen etc. As part of the total over 16 population, the Catholics have 3.4 percent of their men and 2.2 percent of the women in these grades; the 'others' have 5.1 percent and 2.4 percent, and the 'not stateds' 4.9 percent and 2.5 percent. So the differential for men is very high. But if we take these figures only on the employees in employment, the Catholic men have 6.7 percent in these grades, against 8.2 percent of the 'others' and 8.1 percent of the 'not stateds'; this is a much smaller difference. For the non-manual women supervisors etc. the same percentage, roughly, pertains to Catholics in this grade as the percentage of all employed women. Similar considerations apply when we look at the professional grades: 50 percent more Protestants, relatively, were in these grades if we take the whole population, but only about 25 percent more if we reckon the share on the employed population only.

All told, the Greater Belfast employment structure is naturally much more top-weighted than the rest of the country; 19 percent of Catholic male employees are above bottom grade (excluding apprentices), 25 percent of 'others' and 24 percent of 'not stateds'. Thus, where the career opportunities are generally better, Catholics also obtain a higher proportion of jobs above the lowest grade, though this is still not equal to that recorded for the other denominations.

The subdivisions of the travel-to-work area are the districts which compose it. The full results for these have been printed in Additional Table A5.6(a). We here present some of the main findings of this more detailed analysis, partly because it illuminates very clearly the social geography of the Greater Belfast area.

Belfast district itself had almost a quarter of its adults as stated Catholics, about 56 percent as Protestant and 20 percent as 'not stated'. Catholic male unemployment was more than twice as high as for 'others', and for females it was almost twice as great. Self-employment played only a small role, so it matters little whether we take all employed persons, or only employees in employment, as our reduced denominator for looking at the status structure of the labour market. Once again the differential is there, but it shrinks as we exclude the economically non-active and the unemployed. Among non-manual foreman (and women) there is practically no difference – something we attribute to the large number of jobs in the public sector. Moreover, the numbers here are so large that local accident may be excluded from our considerations. There were over 17,000 higher-grade jobs in the Belfast local labour market in 1981, against an economically active population of 132,000, and 103,000 persons in employment.

If we look at the 'not stated' category, it resembles the Catholic population in its status structure as far as small-firm managers are concerned; it resembles the Protestants when we look at professional employees. It is somewhere halfway between when we look at manual foremen, higher than either of the stated denominations among the non-manual male foremen, and similar across the board for the non-manual female supervisors etc. We can draw no conclusions from this as regards the possible composition of this group.

As the outer Belfast group districts mostly constitute large settlements, we have provided an extended analysis (Additional Table A5.6(a)) for all of them. Castlereagh, as we would expect, provides an exceptional structure. Catholic unemployment is actually a lower percentage of the adult male population than is the case for 'others', and a lot better than the 'not stateds'. However, here we are talking about just 66 Catholic males out of about 2,500 unemployed, and a 5 percent share of Catholics in the total over 16 population. If we try to analyse the higher-grade employees, the numbers are again very small. However, for what it is worth, the Catholics here have a disproportionately high share of better jobs: 18 percent of the 614

Catholic male employees were in managerial grades and 9 percent in professional employment, as against 12.7 percent and 5.6 percent respectively of the 'others'. This comes as no surprise: it is a function of the social geography of the Greater Belfast settlement rather than any excess of opportunities.

Much the same may be said about the other two middle-class Belfast suburbs, where Catholic excess unemployment is only moderate (about 50 percent in Newtownabbey and North Down). In Ards the excess is greater: 16.8 percent as against 7.7 percent. In Lisburn, with its very large Catholic housing estates, which are properly part of West Belfast, it is 18.5 percent against 7.1 percent. The 'not stateds' in each case are somewhere halfway between Catholic and Protestant unemployment rates, within totals which are very low by Northern Ireland standards. Only Lisburn and Newtownabbey had more than 10 percent of their economically active male population unemployed.

When we turn to employment status, we see quite a contrast again. North Down (with only 7 percent of its 16 and over population Catholic) has a reasonable structure. Managers and foremen are a little underrepresented, and only professional employees are seriously down on their share – but then we are talking about 48 men and 4 women. Therefore it is in the other three districts that we have to look for the more meaningful differentials, although only in Lisburn is there a sizeable Catholic population. Here the structure of management, foremen and professional grades shows more serious discrepancies: 8 percent of Catholics were managers as against 13.1 percent of 'others' and 12.4 percent of 'not stateds'; among professional employees, 3.6 percent were Catholics against 5.4 percent for 'others' and 5.7 percent of 'not stateds'. Women again attain something nearer parity, at least in the non-manual supervisory grades.

Newtownabbey, with 8 percent Catholic adults, again achieves a more equitable distribution, except for male manual foremen. Among the non-manual foremen grades, Catholics seem slightly favoured – until we see that we are dealing with just 100 men and women, so the evidence does not stand up. Lastly Ards, with under 10 percent Catholic population, and fairly low overall proportions of higher-grade employees, shows a seriously weak structure among the Catholics compared with the 'others' as well as the 'not stateds'.

It seems to us fairly meaningless, in fact, to take the analysis down to this level if the total Catholic share of the local population is only 10 percent or less, and when travel-to-work patterns are as complex as they are in the Belfast area.

Overall we conclude that Greater Belfast, which in any case has the lowest unemployment figures and the best employment status structure, also has relatively fewer Catholics in the lowest economic status categories. Only in Belfast city is Catholic unemployment both a serious and a large problem. A quarter of the city's 25,000 adult males were unemployed, and only 11.5 percent of the 'others'. That is a very large failure indeed. It is less serious for women both absolutely and relatively. But in Belfast the status structure, for those Catholics who are in employment, is less disadvantageous than elsewhere, and the 18 percent of Catholic males in the better jobs represent over 2,200 employees. We have not worked out the much smaller percentages of women, except for the non-manual supervisors etc.; but even there we are talking about over 1,000 women who have, in a sense, made the grade.

Other Area Groups

Detailed analysis of the other travel-to-work areas, and individual districts, reveals some interesting patterns, but no startling differences compared with the national figure. The full results are presented in Additional Table A5.6(a).

In the Northern group, which includes Antrim, a district which is economically more closely allied to Belfast than any other outside the Belfast TTWA, has the smallest proportion of Catholics of the four we have analysed. Unemployment among the Catholic population, male and female, is twice as great as that for the 'others', and nearly the same as for the 'not stateds'. This does not mean that we can be sure that this last group consists mainly of Catholics, since in some other categories, for example students and 'other employees', the 'not stated' group is more like the Protestants and 'others'. The key groups of managers etc. are so small that interdenominational comparisons of percentages are fairly meaningless.

At the level of individual districts, there are no marked variations. In some of the coastal areas (Larne, Moyle and Coleraine) there are relatively high proportions of retired people, presumably mostly in owner-occupied dwellings, and this may explain the relatively high proportion of Protestants and 'others' in the non-employed group. Coleraine has a relatively smaller excess of unemployed Catholics, and also about the same proportion of employees in the 'other' categories, that is, not basic grades, mainly due to a higher proportion of professional workers. Whether these features are related to Coleraine's status as a university town, cannot be stated with certainty: the area certainly stands in marked contrast to other predominantly Protestant settlements of that size.

The Southern area group, in contrast, has a high proportion of Catholics. The differential in unemployment is greater still than in the Northern area: Catholics showed 23.6 percent unemployment for men, as opposed to 8.8 percent for the 'others' and 22.8 percent for the 'not stateds'. This makes the 'not stated' category much more like the Catholics, and this is then also confirmed when we look at other characteristics: percentages in the lower employment status grades and in professional occupations. So the theory that most of the 'not stateds' were Catholics seems to be borne out there. On the other hand, in the 'student' classification, the Catholics have a higher proportion than the 'not stateds'.

At the district level, there are few remarkable features. Newry and Mourne is an area with rather few higher-status occupational groups than the provincial average. Within that relatively narrow stratum of more skilled or qualified workers, the Catholics did not have a noticeably lower share: in fact 13.5 percent of them were employed above the bottom grades in the hierarchy, but only 9 percent of the 'not stateds'. (This once again shows the impossibility of dividing up those who did not answer the religion question according to some standard formula.) However, these statements should all be seen against the background of the fact that out of nearly 50,000 people in Newry and Mourne above the age of 16, only about a third were employees at all; and of these 16,000, only 1,850 were anything more than basic-grade workers, that is 3.5 percent of the adult population.

Of the other districts, two with large Catholic populations, Cookstown and Down, have dissimilar structures. Cookstown is one of the worst areas in the Province, with a third of Catholic men unemployed, whereas Down has only half that percentage. In Cookstown there are practically no professional or managerial workers; in Down there are relatively and absolutely a great many more of these, and the Protestant/Catholic differential is much smaller. Looking at the map more closely, we realize that the Saintfield area of Down is an outlier of Greater Belfast, and seems to be a preferred location for middle-class commuters into central Belfast.

Lastly we look at the Western group, which has a higher proportion of Catholics than any of the others, at 46 percent of all adults. Unemployment is once again sharply differentiated by religion: a quarter of all Catholics, the same for the 'not stateds', and 11 percent of the Protestants and 'others'. The management structure favours the 'others': there would be more Protestant than Catholic managers of large enterprises even if *all* 'not stateds' were Catholics, which is unlikely. Only among the manual male foremen and non-manual female supervisors is there anything like parity in the religious structure of these above basic-grade categories.

Derry is the largest labour market outside Belfast, and has been analysed in some detail. For males, 29 percent of Catholics were unemployed, as against 11.5 percent of 'others' and 26.9 percent of the 'not stateds'. Again for males, 5.5 percent of Catholics were managers and 8.8 percent of 'others'. In the other non-basic grades, Catholic men were fairly favourably placed (that is, had a proportional share of the better jobs), but this was not true for women. So, even in an area allegedly under the control of Catholic employers, the employment structure does not favour Catholics. Professional employees were disproportionately of other denominations.

Of the other areas in the Western group, few show any striking departures from the average. Fermanagh has a higher proportion of retired Catholics; the professional and managerial cadres in this rural district are too small to be analysed in any meaningful way. The districts of Limavady and Strabane are notorious for their high unemployment: nearly 30 percent for Catholic males. The differential against the Protestants and 'other' group is greatest in Omagh, where Catholic rates are 233 percent of Protestant rates; in Strabane the rate is 'only' 75 percent higher. It is very doubtful, however, whether these differences have any real meaning in such small districts which are heavily disadvantaged on all counts. Their industries are disappearing, and what employment exists is largely unskilled. The whole of Strabane district has about 100 professional employees (as against 500 in Derry).

8
Occupational Orders and Religion

We must now look at the position of the Catholics in the labour market in another way. Industrial classifications give one set of clues; the occupational hierarchies used in this chapter afford a rather different set of explanations. Occupational orders indicate not only the kind of product or service provided by the enumerated workers, but also their position within the labour force both according to skill and according to the structure of command and of remuneration. We have already seen, in the preceding chapter, the underrepresentation of Catholics in foreman and manager positions. These imbalances may be the result of a number of factors, ranging from lack of qualifications, through failure to choose an appropriate career, to discrimination at the workplace.

In this chapter we examine the distribution of employed workers between occupations which require different qualifications and afford varying prospects of advancement. They also vary greatly as regards the incomes that can be derived from working in these jobs.

In Table 8.1 we give information about the religious structure of employment in the seventeen occupational orders and some key suborders, by sex. (For list of occupational orders see Appendix C.)

Table 8.1 *Overview of structure of employment by occupational orders and selected suborders of employment, by sex and religion, 1981 (percentages only)*

Occupational	Roman Catholics		All others		Not stated	
order[1]	Male	Female	Male	Female	Male	Female
01	15.8	15.5	67.5	66.2	16.7	18.3
002	17.5	14.5	65.4	66.6	17.0	18.8
005	12.2	10.1	72.4	74.2	14.4	13.7
02	29.4	31.4	53.0	51.7	17.7	16.9
011	34.9	33.2	46.7	51.1	18.4	15.8
016	46.0	33.4	31.9	49.1	22.0	17.5
03	18.4	15.5	56.1	62.0	25.5	22.5
04	14.2	17.2	67.0	62.1	18.8	20.6
030	16.7	18.1	64.2	61.8	19.0	20.1

Table 8.1 *continued*

Occupational order[1]	Roman Catholics Male	Roman Catholics Female	All others Male	All others Female	Not stated Male	Not stated Female
05	21.6	20.1	63.2	64.6	15.1	15.3
038	22.1	18.8	61.8	66.4	16.1	14.7
040	23.6	17.9	62.3	70.0	14.0	12.0
06	20.7	19.8	61.0	63.8	18.3	16.4
046	20.0	21.0	61.9	63.0	18.3	16.0
049		19.3		64.3		16.4
07	19.0	19.0	64.2	65.2	17.0	15.5
055	23.0	20.0	60.0	65.0	18.0	16.0
057	15.0	14.5	68.0	70.0	17.0	15.5
08	11.0	9.0	75.0	76.4	14.0	14.6
058	11.0	7.0	79.0	83.4	10.0	9.6
061	5.0	5.0	77.0	75.8	18.0	18.9
062	16.7	12.8	68.4	72.4	14.8	14.9
09	31.6	25.8	49.4	58.3	19.0	15.8
066		25.8		58.1		15.8
068	33.8	27.2	45.3	57.6	20.8	15.2
072	28.6	21.4	55.9	64.0	15.5	14.6
10	24.3	12.7	60.1	70.3	15.6	16.9
11	27.1	27.8	54.9	55.2	17.8	16.9
098	21.1	19.5	62.0	63.2	16.8	17.3
12	17.9	20.7	65.3	61.3	16.8	18.0
117	16.1	35.6	67.7	47.1	16.2	17.2
13	24.6	19.8	57.9	65.6	17.5	14.6
133	26.3	21.8	56.0	50.0	17.7	28.2
137	18.6	17.9	64.9	67.4	16.4	14.6
138	20.0	22.5	63.0	63.3	16.8	14.2
14	33.8	31.0	47.0	44.4	19.1	24.4
140	34.2	23.8	47.2	52.3	18.6	23.8
143	38.0	40.0	41.4	40.0	20.6	60.0
15	24.1	17.4	58.3	64.1	17.7	18.5
152	24.7	19.3	56.6	61.5	18.6	19.3
157	21.5	16.1	61.9	64.8	16.6	19.1
16	25.2	25.0	57.1	58.0	17.6	17.0
160	25.8	25.2	56.1	57.6	18.0	17.1
17	28.8	32.2	43.6	41.7	27.7	26.1
Total employed	22.3	24.2	60.4	59.2	17.4	16.7

[1] For definition of occupational orders see Appendix C. Percentages are calculated on all employees in the relevant order.

Source: Own calculations based on Northern Ireland Census 1981, *Religion Report*, Table 8 (unpublished information)

We have omitted orders which are relatively insignificant. In some cases we only give males (or females) because they are virtually single-sex occupations. We omit suborder 40 – farmers, horticulturists at local level, and farm managers (within the main order 5, managers) – because their distribution is not relevant to the operation of the employment market (see Chapter 6). The table, for clarity, only gives percentages; for the numbers involved see Additional Table A5.6(b).

The reference point for this table is the position in 1981 of all employed persons, by religion and sex, and this can be summarized as follows:

	Catholic		Others		Not stated	
	M	F	M	F	M	F
Employed (incl. self empl.) as % of all employed (M and F)	22.3	24.2	60.4	59.2	17.4	16.7

It would have been preferable to exclude the self-employed; but the Census tabulations give employment at local level only by total employed persons, so that our percentages have to refer to this larger total. In practice it makes little difference except when it comes to professional and managerial categories, the main classes of self-employment. For example order 5, which includes suborder 40 (farmers etc.), has 17,241 males in the printed economic activity tables but 48,870 in the unpublished and more detailed computer printout of Table 8 (of the religion report), which includes an extra 22,000 self-employed farmers.

The exact differences which arise from the principles of classification are set out in the Census economic activity volume (p. viii), and from this information it is possible to choose the categorization which best meets the needs of an analysis of employment status. Thus in 1981 there were nearly 400,000 economically active males (employed, self-employed or unemployed), as opposed to only 270,000 'employees in employment'. In this chapter we have taken as our basis the employed plus the self-employed, actually working at Census date, because that is the classification which exists in the most detailed form, at district level, by religion. It is unlikely that the results would be very different if one took as the basis, if it were available, a total excluding the self-employed. The larger group of 'economically active', however, would introduce a serious distortion because of the large element of persons whose occupational status was not stated or inadequately described. This would refer largely to the young unemployed who had never been in a stated occupation.

This category alone accounts for about 40,000 of the differences in gross totals of male workers.

As a somewhat arbitrary rule, we shall assume that there will inevitably be some random deviations from national average proportions as set out above, and that it is only worth commenting on the percentages if the Catholics' share of the labour market differs from the average by about 5 percent. Applied to the total Catholic male employment figure the difference would amount to 3,500 workplaces, and applied to females it would be 2,400. (In practice, the numbers will be much smaller in each order.) We immediately see the problem with order 1, the main professional and senior managerial grades (for full list see Appendix C), and with order 2, which includes medical practitioners. In order 1 these grades comprised 12,400 men in 1981. Since Catholic male employees were 22.3 percent of the working population, there should have been 2,765 Catholic men in the specified occupations: in fact there were only 1,956. This appears to show serious underrepresentation. (Readers should ignore published figures in Table 7A of the Census economic activity report, which, as stated, exclude the self-employed.) When we get to order 2 (suborders 11 and 16 in particular), we have the expected high representation of Catholics in the labour force; teachers and nurses provide the great majority. (We note that the overrepresentation of men in the nursing profession is even more noticeable than among the women.) In order 4 (the less senior professional jobs in science and engineering and other technological subjects) there is again serious underrepresentation: 1,700 men instead of 2,700 as expected on a proportionate basis.

Some occupational orders either show no very remarkable deviations, or the numbers involved are very small, until we get to suborder 57, sales staff (that is, representatives etc.), where again we have a serious underrepresentation of Catholic males. The position in order 8 (security and protective services) is entirely as expected. It is now one of the most important branches of employment, with 23,000 men and women employed in 1981. But instead of the expected proportion of 4,700 men we find 2,340, and instead of the expected 440 women there were only 164.

In order 9 (catering, miscellaneous services etc.) we have more than proportionate representation, especially among men, and when we look at the full list we see the descriptions: counter-hands, caretakers, domestic staff, school helpers, porters, sweepers, cleaners. Though the percentage figures are high for men, the numbers involved are much greater for women; 32,500 of them are married women, 28 percent of all working married women. Orders 11 (really consisting of craft occupations which may occur in industrial undertakings or in retail distribution, and for convenience labelled 'industrial'), 12 and 13 relate to the main industrial occupations, and the picture here again is much as expected. The Catholic men are underrepresented in engineering, typically so in suborder 117 (metal working, production fitters etc.). The figure for women looks promising at over a third of the whole labour force, until we check the actual figures: there is a total of 87 women in the country in that category, compared with just under 10,000 men.

Order 14 comprises most of the construction trades, and here Catholics are strongly represented. But again the skill structure is interesting: order 140 is unspecified building and construction workers and order 143 is labourers, and these are the two categories which comprise two-thirds of all building workers.

There is nothing very remarkable about the other orders and suborders, except perhaps for order 17, namely inadequately described occupations and unstated occupations. Here there were 15,000 men and women who said nothing about themselves in the Census which could be analysed, though they were working. If we took this group from the analysis of economically active persons, that is including the unemployed who had never worked, it would amount to 75,000 men and women.

To sum up, the occupational orders analysis bears out what the studies of employment status, industrial structure and socio-economic classifications (Chapters 5–7) have suggested: that Catholics are more than proportionately represented in the low-skill, low-paid, part-time, no-prospect jobs, and underrepresented in those areas where skill, status, pay and prospects are all positively rated. This is no more and no less than common sense and daily observation suggest, but it is as well to be reasonably precise about the matter.

Once again, we need not be unduly concerned about underenumeration. If people escaped the Census net, they were more likely to be on the low-skill side of the spectrum; if they did not state their religion, then the matter is not so clear-cut. So the 79,000 who were not counted would, if included, probably still further weight the balance of the skill structure against the Catholics.

As for the 'not stateds', it is clear from Table 8.1 that they are *not* like the Catholics in many important categories where we can make informed guesses about skills, pay etc. Thus order 2 (teachers and nurses) is much nearer the standard proportion for the 'not stateds' than it is for Catholics, except possibly for male nursing staff (where Catholics had 46 percent of the suborder against the expected 22 percent), and the 'not stateds' have 22 percent instead of 17.4 percent. The 'not stateds' are not seriously underrepresented in the professional and managerial cadres. They are a little low in order 8 (security) and in suborder 72 (caretakers); they have average shares in industrial occupations. Overall, however, the 'not stateds' are remarkably consistent in having in almost all orders the same share which they have nationally, in the total labour force.

We now have to examine, without going into too much detail, the local position, if only to see whether and where there are serious anomalies.

Local Structure

We shall try to see how far the local occupational structure differs from the national picture. In order not to run the danger of making too much of apparently large percentage differences when the absolute numbers are very small, we reduce the number of occupational orders analysed. Percentages are not generally given where the total of employees in a particular suborder involves fewer than 2,000 men and women, except where we are dealing with particularly significant status groups, especially for women.

The data for this section is presented in Table 8.2 and Additional Table 5.6(b). At the end of each area group division in the tables we print the totals of males and females, by denomination, employed in that area, and the percentages they form of the total labour market. These figures are the same as those in the employment status tables, where we measure shares of status groups against total employment (not of employees in employment). These percentages may be used as reference points to determine whether shares in particular occupational orders are, or are not, in proportion to the whole workforce.

Area Group I: Belfast

In the professional and higher managerial echelons (order 1) we have 11.5 percent Catholic males against 15.7 percent of the over 16 population, or against 13.5 percent of the employed population. (Tables 8.2 and A5.6(b) give percentages by order, suborder, gender and denomination against all those in employment.) So once again the differential shrinks somewhat when we take only those who were actually working, as opposed to the whole adult population. The 'others', whose share of the employed population was 68.6 percent, had 71.1 percent of order 1 jobs, and the 'not stateds' had 17.4 percent.

It will be seen that, because of the total numbers involved, a slightly more than proportionate number of 'others' in this order is reflected in a relatively more serious discrepancy for the Catholic population. The 'not stateds' group has about the expected share of jobs in that order, thus giving no clue as to the religious affiliation of

Table 8.2 *Overview of structure of employment by occupational orders and selected suborders of employment, by sex and religion, by area group, 1981 (numbers and percentages)*

Area groups[1] and orders[2]		All denominations Male	All denominations Female	Roman Catholics Male	Roman Catholics Female	All others Male	All others Female	Not stated Male	Not stated Female
I Belfast									
01	(no.)	8,414	1,613	968	210	5,983	1,089	1,463	314
	(%)			11.5	13.0	71.1	67.5	17.4	19.5
02	(no.)	9,433	17,709	1,711	3,681	5,886	10,823	1,836	3,205
	(%)			18.1	20.8	62.4	61.1	19.5	18.1
011	(no.)	3,759	5,873	888	1,258	2,117	3,551	754	1,064
	(%)			23.6	21.4	56.3	60.5	20.1	18.1
016	(no.)	375	8,070	112	1,818	174	4,828	89	1,424
	(%)			29.9	22.5	46.4	59.9	23.7	17.6
04	(no.)	7,687	766	766	111	5,419	476	1,502	179
	(%)			10.0	14.5	70.5	62.1	19.5	23.4
05	(no.)	15,693	3,625	1,673	444	11,523	2,629	2,497	552
	(%)			10.7	12.2	73.4	75.5	15.9	15.2
038	(no.)	5,317	1,728	624	187	3,795	1,290	898	251
	(%)			11.7	10.8	71.4	74.7	16.9	14.5
06	(no.)	14,329	33,656	2,125	4,738	9,420	23,142	2,784	5,776
	(%)			14.8	14.1	65.7	68.9	19.4	17.2
046	(no.)	9,324	18,831	1,274	2,742	6,248	12,926	1,802	3,163
	(%)			13.7	14.6	67.0	68.6	19.3	16.8
049	(no.)		8,881		1,171		6,166		1,544
	(%)				13.2		69.4		17.4
07	(no.)	7,425	8,975	790	1,085	5,307	6,470	1,328	1,420
	(%)			10.6	12.1	71.5	72.1	17.9	15.8

Area groups[1] and orders[2]		All denominations		Roman Catholics		All others		Not stated	
		Male	Female	Male	Female	Male	Female	Male	Female
055	(no.)	1,693	8,012	229	979	1,145	5,768	319	1,265
	(%)			13.5	12.2	67.6	72.0	18.8	15.8
057	(no.)	4,614	574	422	58	3,366	424	826	92
	(%)			9.1	10.1	73.0	73.9	17.9	16.0
08	(no.)	9,816	986	773	64	7,483	773	1,560	149
	(%)			7.9	6.5	76.2	78.4	15.9	15.1
09	(no.)	5,017	22,601	1,236	3,864	2,789	15,200	992	3,537
	(%)			24.6	17.1	55.6	67.3	19.8	15.6
066	(no.)		3,582		602		2,401		579
	(%)				16.8		67.0		16.2
068	(no.)	55	7,172	13	1,177	28	4,915	14	1,080
	(%)			23.6	16.4	50.9	68.5	25.5	15.1
070	(no.)	294	1,130	65	254	179	696	50	180
	(%)			22.1	22.5	60.9	61.6	17.0	15.9
072	(no.)	1,769	5,662	327	855	1,156	4,022	286	785
	(%)			18.5	15.1	65.3	71.0	16.2	13.9
10	(no.)	1,837	176	206	9	1,366	132	265	35
	(%)			11.2	5.1	74.4	75.0	14.4	19.9
11	(no.)	11,947	6,695	1,667	911	8,187	4,736	2,093	1,048
	(%)			14.0	13.6	68.5	70.7	17.5	15.7
12	(no.)	27,698	1,120	2,801	85	20,254	845	4,643	190
	(%)			10.1	7.6	73.1	75.4	16.8	17.0
13	(no.)	4,709	2,619	835	215	3,067	2,037	807	367
	(%)			17.7	8.2	65.1	77.8	17.1	14.0

Table 8.2 *continued*

Area groups[1] and orders[2]		All denominations		Roman Catholics		All others		Not stated	
		Male	Female	Male	Female	Male	Female	Male	Female
14	(no.)	5,768	21	1,195	1	3,539	12	1,034	8
	(%)			20.7	4.8	61.4	57.1	17.9	38.1
15	(no.)	13,476	447	2,228	53	8,924	315	2,324	79
	(%)			16.5	11.9	66.2	70.5	17.2	17.7
152	(no.)	6,223		1,046		4,010		1,167	
	(%)			16.8		64.4		18.8	
157	(no.)	4,151		689		2,785		677	
	(%)			16.6		67.1		16.3	
16	(no.)	5,098	358	804	50	3,469	262	825	46
	(%)			15.8	14.0	68.0	73.2	16.2	12.8
160	(no.)	4,714		758		3,181		775	
	(%)			16.1		67.5		16.4	
17	(no.)	4,178	2,857	844	663	2,168	1,433	1,166	761
	(%)			20.2	23.2	51.9	50.2	27.9	26.6
Total in	(no.)	153,595	104,710	20,720	16,244	105,434	70,679	27,441	17,787
employment	(%)			13.5	15.5	68.6	67.5	17.9	17.0
II Northern									
01	(no.)	1,370	268	175	26	1,008	201	187	41
	(%)			12.8	9.7	73.6	75.0	13.6	15.3
02	(no.)	3,037	6,100	734	1,714	1,848	3,454	455	932
	(%)			24.2	28.1	60.8	56.6	15.0	15.3
011	(no.)	1,313	2,362	372	679	742	1,366	199	317
	(%)			28.3	28.7	56.5	57.8	15.2	13.4
016	(no.)	267	2,900	84	892	131	1,536	52	472
	(%)			31.5	30.8	49.1	53.0	19.5	16.3

Area groups[1] and orders[2]		All denominations		Roman Catholics		All others		Not stated	
		Male	Female	Male	Female	Male	Female	Male	Female
04	(no.)	1,825	132	250	16	1,277	96	298	20
	(%)			13.7	12.1	70.0	72.2	16.3	15.1
05	(no.)	9,608	1,265	1,694	230	6,674	840	1,240	195
	(%)			17.6	18.2	69.5	66.4	12.9	15.4
038	(no.)	1,869	578	354	89	1,267	395	248	94
	(%)			18.9	15.4	67.8	68.3	13.3	16.3
06	(no.)	2,602	7,388	460	1,194	1,763	5,126	379	1,068
	(%)			17.7	16.2	67.8	69.4	14.6	14.5
046	(no.)	1,679	4,434	301	713	1,130	3,045	248	676
	(%)			17.9	16.1	67.3	68.7	14.8	15.2
049	(no.)		1,678		282		1,175		221
	(%)				16.8		70.0		13.2
07	(no.)	1,942	2,364	290	386	1,378	1,660	274	318
	(%)			14.9	16.3	71.0	70.2	14.1	13.5
055	(no.)	577	2,168	89	360	409	1,519	79	289
	(%)			15.4	16.6	70.9	70.1	13.7	13.3
08	(no.)	3,418	305	374	30	2,607	234	437	41
	(%)			10.9	9.8	76.3	77.4	12.8	13.4
09	(no.)	1,443	6,492	352	1,384	837	4,112	254	996
	(%)			24.4	21.3	58.0	63.3	17.6	15.3
066	(no.)		1,185		233		794		158
	(%)				19.7		67.0		13.3
068	(no.)	19	2,038	7	463	7	1,277	5	298
	(%)			36.8	22.7	36.8	62.7	26.4	14.6
072	(no.)	519	1,342	114	239	340	893	65	210
	(%)			22.0	17.8	65.5	66.5	12.5	15.6

Table 8.2 *continued*

Area groups[1] and orders[2]		All denominations		Roman Catholics		All others		Not stated	
		Male	Female	Male	Female	Male	Female	Male	Female
10	(no.)	1,343	201	280	13	887	155	176	33
	(%)			20.8	6.5	66.0	77.1	13.1	16.4
11	(no.)	6,250	3,263	1,255	657	3,936	2,036	1,059	570
	(%)			20.1	20.1	63.0	62.4	16.9	17.5
12	(no.)	7,647	642	1,316	106	5,156	416	1,175	120
	(%)			17.2	16.5	67.4	64.8	15.4	18.7
13	(no.)	1,073	856	206	141	673	604	194	111
	(%)			19.2	16.5	62.7	70.6	18.1	13.0
14	(no.)	3,186	9	854	5	1,793	4	539	—
	(%)			26.8	55.6	56.3	44.4	16.9	0
15	(no.)	4,986	99	930	15	3,229	66	827	18
	(%)			18.7	15.2	64.8	66.7	16.6	18.2
152	(no.)	2,656		514		1,695		447	
	(%)			19.4		63.8		16.8	
157	(no.)	1,077		157		740		180	
	(%)			14.6		68.7		16.7	
16	(no.)	1,788	213	318	32	1,143	135	327	46
	(%)			17.8	15.0	63.9	63.4	18.3	21.6
160	(no.)	1,652		300		1,041		311	
	(%)			18.2		63.0		18.8	
17	(no.)	1,482	975	348	241	768	492	366	242
	(%)			23.5	24.7	51.8	50.5	24.7	24.8
Total in employment	(no.)	53,200	30,635	9,875	6,200	35,098	19,678	8,227	4,757
	(%)			18.6	21.5	66.0	64.2	15.5	16.5

Area groups[1] and orders[2]		All denominations		Roman Catholics		All others		Not stated	
		Male	Female	Male	Female	Male	Female	Male	Female
III Southern									
01	(no.)	1,755	324	495	84	984	192	276	48
	(%)			28.2	25.9	56.1	59.3	15.7	14.8
02	(no.)	4,113	8,639	1,826	3,821	1,611	3,416	676	1,402
	(%)			44.4	44.2	39.2	39.5	16.4	16.2
011	(no.)	1,884	3,307	905	1,499	627	1,322	352	486
	(%)			48.0	45.3	33.3	40.0	18.7	14.7
016	(no.)	555	4,297	335	1,972	98	1,550	122	775
	(%)			60.4	45.9	17.7	36.1	22.0	18.0
04	(no.)	1,862	180	423	51	1,099	101	340	28
	(%)			22.7	28.3	59.0	56.1	18.3	15.6
05	(no.)	14,036	1,718	3,932	475	7,851	976	2,253	267
	(%)			28.0	27.6	55.9	56.8	16.1	15.5
038	(no.)	2,887	781	932	215	1,478	454	477	112
	(%)			32.3	27.5	51.2	58.1	16.5	14.3
06	(no.)	3,983	9,810	1,211	3,092	2,053	5,127	719	1,591
	(%)			30.4	31.5	51.5	52.3	18.1	16.2
046	(no.)	2,645	6,155	797	1,986	1,366	3,181	482	988
	(%)			30.1	32.3	51.6	51.7	18.2	16.1
049	(no.)		2,360		755		1,214		391
	(%)				32.0		51.4		16.6
07	(no.)	2,742	2,745	815	831	1,459	1,459	468	455
	(%)			29.7	30.3	53.2	53.2	17.1	16.6
055	(no.)	823	2,577	261	785	413	1,368	149	424
	(%)			31.7	30.5	50.2	53.1	18.1	16.5
057	(no.)	1,351	86	364	23	785	52	202	11
	(%)			26.9	26.7	58.1	60.5	15.0	12.8

Table 8.2 *continued*

Area groups[1] and orders[2]		All denominations		Roman Catholics		All others		Not stated	
		Male	Female	Male	Female	Male	Female	Male	Female
08	(no.)	3,894	317	531	44	2,814	223	549	50
	(%)			13.6	13.9	72.3	70.3	14.1	15.8
09	(no.)	1,926	8,883	801	3,475	774	3,893	351	1,515
	(%)			41.6	39.1	40.2	43.8	18.2	17.1
066	(no.)		1,415		549		625		241
	(%)				38.8		44.2		17.0
068	(no.)	38	3,507	16	1,448	15	1,491	7	568
	(%)			42.1	41.3	39.5	42.5	18.4	16.2
072	(no.)	737	1,478	290	512	335	724	112	242
	(%)			39.3	34.6	45.5	49.0	15.2	16.4
10	(no.)	2,103	178	695	37	1,034	115	374	26
	(%)			33.0	20.8	49.2	64.6	17.8	14.6
11	(no.)	8,174	5,109	3,052	1,664	3,602	2,604	1,520	841
	(%)			37.3	32.6	44.1	51.0	18.6	16.5
12	(no.)	7,858	362	2,338	118	4,164	171	1,356	73
	(%)			29.8	32.6	53.0	47.2	17.3	20.2
13	(no.)	1,485	1,310	527	329	685	785	273	196
	(%)			35.5	25.1	46.1	59.9	18.4	15.0
14	(no.)	5,148	10	2,250	5	1,822	2	1,076	3
	(%)			43.7	50.0	35.4	20.0	20.9	30.0
15	(no.)	6,230	88	2,090	25	2,923	46	1,217	17
	(%)			33.5	28.4	46.9	52.3	19.5	19.3
152	(no.)	3,644		1,207		1,706		731	
	(%)			33.1		46.8		20.1	

Area groups[1] and orders[2]		All denominations		Roman Catholics		All others		Not stated	
		Male	Female	Male	Female	Male	Female	Male	Female
157	(no.)	1,134		338		589		207	
	(%)			29.8		51.9		18.3	
16	(no.)	2,183	200	906	68	849	100	428	32
	(%)			41.5	34.0	38.9	50.0	19.6	16.0
160	(no.)	2,055		877		768		410	
	(%)			42.7		37.4		20.0	
17	(no.)	2,236	1,359	832	564	748	435	656	360
	(%)			37.2	41.5	33.5	32.0	29.3	26.5
Total in employment	(no.)	70,021	41,326	22,820	14,708	34,619	19,694	12,582	6,924
	(%)			32.6	35.6	49.4	47.7	18.0	16.8
IV Western									
01	(no.)	874	148	318	45	409	75	147	28
	(%)			36.4	30.4	46.8	50.7	16.9	18.9
02	(no.)	2,837	4,957	1,436	2,527	940	1,662	461	768
	(%)			50.6	51.0	33.1	33.5	16.2	15.5
011	(no.)	1,238	2,154	697	1,107	341	755	200	292
	(%)			56.3	51.4	27.5	35.1	16.2	13.6
016	(no.)	392	2,152	200	1,142	104	641	88	369
	(%)			51.0	53.1	26.5	29.8	22.4	17.1
04	(no.)	1,043	105	326	26	525	62	192	17
	(%)			31.3	24.8	50.3	59.0	18.4	16.2
05	(no.)	9,533	950	3,286	372	4,824	436	1,423	142
	(%)			34.5	39.2	50.6	45.9	14.9	14.9
038	(no.)	1,620	395	671	165	684	174	265	56
	(%)			41.4	41.8	42.2	44.1	16.4	14.2

Table 8.2 *continued*

Area groups[1] and orders[2]		All denominations		Roman Catholics		All others		Not stated	
		Male	Female	Male	Female	Male	Female	Male	Female
06	(no.)	2,533	5,574	1,060	2,173	1,065	2,526	408	875
	(%)			41.8	39.0	42.0	45.3	16.1	15.7
046	(no.)	1,726	3,426	708	1,303	741	1,586	277	537
	(%)			41.0	38.0	42.9	46.3	16.0	15.7
049	(no.)		1,353		554		589		210
	(%)				40.9		43.5		15.5
07	(no.)	1,628	1,793	687	750	676	768	265	275
	(%)			42.2	41.8	41.5	42.8	16.3	15.3
055	(no.)	606	1,652	263	687	241	706	102	259
	(%)			43.4	41.6	39.8	42.7	16.8	15.7
08	(no.)	4,136	213	662	26	3,072	162	402	25
	(%)			16.0	12.2	74.3	76.1	9.7	11.7
09	(no.)	1,565	6,040	756	2,659	524	2,459	285	922
	(%)			4.8	44.0	3.3	40.7	1.8	15.3
066	(no.)		1,208		525		499		184
	(%)				43.5		41.3		15.2
068	(no.)	27	2,115	11	946	13	865	3	304
	(%)			40.7	44.7	48.1	40.9	11.1	14.4
072	(no.)	586	998	302	423	187	431	97	144
	(%)			51.5	42.4	31.9	43.2	16.6	14.4
10	(no.)	1,213	65	395	20	619	34	199	11
	(%)			32.6	30.8	51.0	52.3	16.4	16.9
11	(no.)	4,710	3,303	2,467	1,890	1,355	769	888	644
	(%)			52.4	57.2	28.8	23.3	18.9	19.5

Area groups[1] and orders[2]		All denominations		Roman Catholics		All others		Not stated	
		Male	Female	Male	Female	Male	Female	Male	Female
12	(no.)	5,016	328	2,173	198	1,916	71	927	59
	(%)			43.3	60.4	38.2	21.6	18.5	18.0
13	(no.)	827	845	421	431	260	265	146	149
	(%)			50.9	51.0	31.4	31.4	17.7	17.6
14	(no.)	2,916	5	1,461	3	852	2	603	—
	(%)			50.1	60.0	29.2	40.0	20.7	0
15	(no.)	3,689	63	1,581	28	1,461	20	647	15
	(%)			42.9	44.4	39.6	31.7	17.5	23.8
152	(no.)	2,140		861		887		392	
	(%)			40.2		41.4		18.3	
16	(no.)	1,170	129	556	75	387	25	227	29
	(%)			47.5	58.1	33.1	19.4	19.4	22.5
160	(no.)	1,071		514		340		217	
	(%)			48.0		31.7		20.3	
17	(no.)	1,455	956	665	513	390	204	400	239
	(%)			45.7	53.7	26.8	21.3	27.5	25.0
Total in employment	(no.)	45,299	25,508	18,333	11,746	19,320	9,559	7,646	4,203
	(%)			40.5	46.0	42.6	37.5	16.9	16.5

[1] For definition of area groups see Appendix A.
[2] For definition of occupational orders see Appendix C. Lower figures give percentages of people in these groups against all men and women in that order.

Source: Own calculations based on Northern Ireland Census 1981, *Religion Report* tables (unpublished information)

that particular category. We also note that the discrepancies are not as great as they were in the employment status tables in the previous chapter. This is undoubtedly due to the fact that the occupational order statistics include the self-employed, a category strongly represented in order 1, whereas the industry tables in Chapter 7 only include dependent employees. The clear conclusion is that Catholics stand a better chance of rising in the social hierarchy if they are not subject to the rules governing selection for the higher echelons in the larger industrial and commercial enterprises, and in the public service. (These appear mainly in orders 4 and 5.)

In order 2 we get the expected higher percentages, once again especially so for male nursing staff etc. In orders 4 and 5 (the remaining professional and managerial grades) we see the same relatively low percentages except in suborder 44 (all other managers), which is not, however, a significant proportion of all managerial jobs in the area group.

In order 8 (security) we have the expected underrepresentation, though perhaps not as much: 986 male jobs against the 1,315 which would be proportionate to all employment. However, this is not the full picture: if we take out suborder 60 (police sergeants and other supervisory grades in the whole group of security and related services, including fire services etc.) the situation is different and perhaps a little unexpected. Nationally, the Catholics furnish 11 percent of all employed persons in order 8 and only 8.1 percent of the supervisory grades in suborder 60; in Belfast (city district only) they form 12.1 percent of order 8 and 11.7 percent of the higher ranks in the total force. Overall, of course, Catholic men in Belfast constitute 21.5 percent of the employed population, so that the total number of jobs in the order is only about half that which would be proportional; but it is interesting to note that the lack of chances of promotion, on these figures, seems to be less serious in Belfast. We shall not pursue these matters in detail in what follows: they have been adequately dealt with in the specialized field investigations undertaken by the FEA and others, and it is not always possible, without knowing the exact composition of the labour force (age, education, training) to make detailed statements about the share of the denominations in the labour market at the disaggregated level.

We have the usual more than proportionate representation of Catholics in order 9 (catering etc.), though again we have to caution against reading too much into a figure like 23.6 percent in suborder 68, where we have a grand total of 55 male domestic staff and school helpers.

Orders 11, 12, and 13 (the main manufacturing industry occupations) are much more interesting, accounting as they do for 44,000 male and over 10,000 female jobs. We now see quite a sharp difference in the mainstream engineering jobs (order 12), where Catholics are seriously underrepresented on the male side and even more so on the female side. Of the 27,700 jobs in that general order, 11,700 were in Belfast district, and pro rata the same underrepresentation occurs there (see Additional Table 5.6(b)). However, since this is an occupational and not an industrial analysis, we cannot get at the root of the matter without looking at the most disaggregated (and unpublished) analysis of the suborders. An example is suborder 126 (sheet metal workers, platers, shipwrights, riveters etc.) where nationally there are 3,080 male jobs of which Catholics had 320. We look at Belfast district figures and find 1,278 men in these trades, 42 percent of the national total. Of these, 93 were Catholics, or 7.3 percent. This again is an obvious reflection of the situation in the shipyard. So it is worth pointing out that there are other skilled trades in order 12 where Catholics are not quite so much underrepresented, or even account for more than the average 13.4 percent of all employed people. Thus in suborder 117 (metal working production fitters) the Catholic share in Belfast is 13.5 percent. Once again we recall that in Belfast district Catholic males constitute 21.5 percent of the workforce. In suborder 118 (motor vehicle and aircraft mechanics) the national share of Catholic males is 21.4 percent and in Belfast it is 20.7 percent, something nearer the 'correct' share. The telephone authorities seemed to employ Catholics in skilled grades: 21.8 percent nationally and 24.6 percent in Belfast district (suborder 122).

Thus the rather bare bones of analysis by whole orders and area groups hides considerable local variations. When we analysed the industrial structure, as opposed to the occupational breakdown, we were able to compare employment by industry with the Census of Employment statistics for 1981, which listed the number of establishments in each category. Our impression was that the smaller the mean number of employees in any given category of manufacturing industry, in the Census of Employment, the more nearly proportionate the share of the Catholic employees. But this theory would require extensive testing.

Order 13 comprises a miscellany of industrial occupations outside the skilled metal and electrical trades. There are fewer employees here, the size of firms is likely to be very much smaller, and the proportion of Catholic males employed is rather higher, though it is still poor for women (only 215 of the latter are involved in area group I). This turns out to be due mainly to two groups of trades: 133, painters and decorators, and 135, the group described as 'repetitive assemblers' (of metal and electrical goods), where the national share was 33.9 percent and the Belfast district share 45.9 percent for men. It hardly needs saying that these are the least desirable jobs in manufacturing industry, because they are the worst paid and the least secure. (In contrast to the gender composition of that order in the rest of Britain, women do not play such a large role in this suborder in Northern Ireland.)

We must now return for a moment to order 11, the least industrialized of the manufacturing group of occupations. This order comprises the butchers, bakers and candlestick makers, as it were; that is, the trades where small establishments prevail (except in chemicals, where there are large firms). In this group Catholic males are not noticeably underrepresented, and the women also have about their proportionate share. This does not apply to the suborders equally. For instance, in order 93 (foremen in textile processing), among print trade foremen, printing workers and a few others, there is underrepresentation nationally; but there are apparently higher shares among the chemical foremen, as there are among butchers and bakers, and woodworkers. We cannot tell how far this is due to the prevalence of community-based businesses or industries with a high location coefficient in some areas. In Belfast district, the men in order 11 have just a little less than their expected share; they are seriously underrepresented in the printing trades, and the women do badly in the Belfast textile industries, though they are somewhat better represented nationally. (We shall see why when we look at area group IV, the West.)

Thus, overall, the position in the manufacturing- and craft-based industries in Greater Belfast is very uneven, ranging from about adequate Catholic shares in some sections of the labour market to very low shares indeed in the shipyard. It seems, however, that the position in some of the smaller industries and smaller establishments needs to be looked at in some detail to get a general overview. Does this differential point the way to policy prescriptions?

The remaining industrial orders require little comment. Construction workers are again well to the fore among the Catholics, with 20.7 percent of the male employed being of that denomination. But as in the national picture, when we look at the skill structure the picture is not so rosy. In Belfast, the Catholic men in suborders 140 and 143 (general workers and labourers) account for 70 percent of the workforce. In suborder 140 we find 32.6 percent Catholics and in suborder 143 we have 27.7 percent, but in suborder 139 (foremen in building and civil engineering) this percentage is 26.8 and among the water and sewerage services group it is 17.5. So the Belfast total building and construction workers' labour force of 2,223 men contains just over 30 percent Catholics, but that turns out to be almost entirely due to the large body of relatively unskilled workers – with the share of foremen being a bright spot.

The last three main industrial orders (14, 15 and 16) have practically no women among them, so we have not analysed them.

In Additional Table A5.6(b) we reproduce the district figures for other places in area group I besides Belfast. These invariably consist of rather small cells of employees in the different orders and suborders, so that a more disaggregated analysis would lead to spurious precision about the operation of the labour markets.

Castlereagh's social structure leads us to expect relatively large numbers in orders 1, 2, 4, 5, 6 and 7, and that is what we get, with better than expected proportions only in order 1 (the senior professional and managerial and civil service grades). There is the usual order 2 excess share. On inspection of the detailed printout, this assumption proves correct. Castlereagh district houses 279 of the country's 2,400 national government officers (male plus female). Of these nationally 13 percent are Catholics, but only 8 percent of those resident in Castlereagh; but since this is just about double the proportion of Catholic employed males in Castlereagh, we account for the high proportion in order 1.

By way of contrast, we can take Lisburn. This has a Catholic percentage of 13 for employed males and 14 for women. If we look at the breakdown for the district, we get a distinct underrepresentation for all the upper-echelon orders and the usual excess of shares for order 2. Order 8 is well under average, and order 9 well above. Manufacturing industry shows up quite well in Lisburn, but we remember that we are talking about 1981, and this affects the issue quite clearly: one wonders how many of the 3,000 jobs in order 12 (which includes most of the skilled and assembly workers in the motor vehicle industry) survived the closure of the De Lorean plant. The industrial outlook for Catholic women was particularly favourable in Lisburn. The remaining orders in that district were either numerically insignificant or showed no remarkable deviation from the standard distribution.

For the other highly Protestant middle-class district of North Down, we have a rather similar picture to Castlereagh but without the apparent high representation of Catholics in the higher grades. As a small anomaly, we note that Catholics in the security services are marginally better represented in North Down than their overall share of residents would suggest: this might be interpreted as meaning that Catholics who do work in these occupations prefer to live in an environment like that provided by North Down rather than to stay in areas where Catholics predominate. Closer inspection of the original records in fact reveals that the men in question are mostly members of the UK military forces, presumably stationed in barracks on the Belfast borders, and so we can dispose of the anomaly. The point is only mentioned here because it shows how useless even a detailed local analysis could be without a precise knowledge of the local settlement patterns. Yet, as we know, Northern Ireland is not a single labour market; there are considerable obstacles to the crossing of certain boundaries, and it is therefore inevitable that distortions of this kind should arise. They do give us a clue to the way the labour market fails to produce any sort of equilibrium between residents and work opportunities, even within a general framework of high unemployment.

Area Group II: Northern

In the Northern group of districts, with a total of 81,000 people in employment, we have 18.6 percent of the males in employment stated as Catholics, and 21.5 percent of the women. It is at once apparent that, when we get away from Belfast, the share of Catholics in the better jobs drops even further. Orders 1, 4 and 5 (the main professional and managerial grades) all show quite a serious deficit against the expected proportion, except for suborder 38, who are mainly shop managers. In other selling occupations their share is also low; otherwise the national pattern is repeated. Industrial occupations are (or were) perhaps less distorted than in Belfast, except for women in the larger manufacturing industries; however, there is an exception there in the case of process and assembly workers, where Catholic women do not seem to have secured the same share of jobs as, for instance, in the craft-based industries.

As elsewhere, Catholics are highly represented in the miscellaneous order 17 (inadequately described and not stated), and unsurprisingly this order also figures more largely among the 'not stateds'.

Among the districts in the Northern group, Antrim, Ballymena and Coleraine all had significant concentrations of manufacturing industry. In the first of these, the Catholic share of industrial occupations was not noticeably low; in Ballymena and Coleraine, however, there was a serious underrepresentation of Catholic workers in these fields, and the 'not stateds' equally fell well below their proportionate share. Only in craft-based industries in Coleraine was there a better share-out of jobs. The other districts had so little manufacturing industry that any analysis becomes too hazardous.

In Antrim we had hoped to identify the airport workers, especially in the higher echelons (pilots, flight controllers and airport managers), but unfortunately so few were found among those resident in the district that we could not say what influence the airport had on the professional and managerial structure, if any.

However, within the whole area group II, both the transport industries in general and the managerial grades in particular show a very similar underrepresentation of Catholics. Only in Ballymoney, where Catholics formed a slightly higher proportion of the population than in the northern areas generally (though not as high as in Magherafelt or Moyle), was there a relatively equable share of jobs in the different parts of the occupational hierarchy, always excepting order 1. However, since the district was rather deficient altogether in that type of job, we cannot make too much of this effect: seven more Catholics in these better jobs would have 'corrected' the balance.

Area Group III: Southern

We now come to the Southern group. This is known to have a much higher proportion of Catholics but a fairly weak industrial structure outside Craigavon, which accounted for a third of the area's jobs in manufacturing industry, and more than that in the case of women workers.

In this area group, underrepresentation in the professional and managerial cadres was not as serious as elsewhere for either men or women. On the other hand, the excess of Catholic women in the low-grade service occupations is reproduced here as elsewhere, and of course the further we get away from manufacturing industry the more significant these jobs are as part of the total available workplaces. For example, in Craigavon 2,740 women's jobs were in manufacturing and 2,158 in the main service category, order 9; in Newry and Mourne just over 800 were in manufacturing and 1,628 in order 9; in Armagh there were 750 in manufacturing and 1,282 in order 9; and so on.

Craigavon shows poor Catholic representation in all professional and managerial occupations, except again for shop managers, and even in the lower clerical grades. In production industry, we again get the better showing in the case of the craft-based occupations, an inadequate share of jobs for Catholic men in the metal processing jobs, but a more reasonable representation in the non-metal assembling, painting and packaging occupations.

Newry and Mourne is the most heavily Catholic area of the south; and it has all the adverse socio-economic structural characteristics of the peripheral areas. However, we now see that within that structure, Catholic men and women seem to have secured a better share of professional and managerial jobs; they are well represented in the clerical occupations and, interestingly enough, their share of the low-grade service jobs (in order 9) is not all that excessive, certainly not for women. All told, in Newry and Mourne, with the obvious exception of the security operations, and also except among farm employees, the Catholic occupational structure seems to reflect the composition of the workforce in general. The 'not stateds' form a high percentage of the total, but their occupational structure follows no consistent pattern, and from that evidence alone we could not even guess what proportion of them were Catholics.

Area Group IV: Western

We now complete the picture by looking at the Western districts – disadvantaged in every respect. Of the 15,000 remaining production and industrial jobs, 6,500 were in Derry, and only here did male jobs in manufacturing proper exceed the number in craft production trades.

Unlike Newry and Mourne, area group IV altogether and Derry in particular, despite the heavy Catholic majority, did not produce a proportionate share of professional and senior managerial jobs for them, and even Catholic female clerical workers were underrepresented. This is particularly noticeable in Derry: 46.5 percent of the males in employment were stated Catholics, but only 36.4 percent of professional cadres, and 39.7 percent of the less senior managers. Catholics were not overrepresented in the low-grade order 9 occupations, and they had a better than expected share of production jobs, most of all in the craft-based occupational orders.

All this has to be seen against the background of overall very high unemployment: at least those Catholics who did get jobs were not confined to the bottom jobs. What few posts there were in senior professional and managerial categories, including local government, did not work out too much to the detriment of Catholics: a higher than expected percentage went to Protestants, but again numbers are so small that a shift of ten officials could produce a very different result. We must bear in mind the problem of age structure: senior employees would be mostly aged over 35, and any reforms in educational curricula and employment practices after 1965 would hardly show up in the pattern of 1981 (see Chapter 9).

In the smaller Western districts we deal once again with very small groups of those workers in whom we must take special interest. Within a weak industrial structure, Catholics everywhere secured jobs mainly in the craft occupations outside metal and engineering and other processing industries. The women did not even get an adequate number of junior clerical jobs. The picture seems much the same (allowing for small numbers) in the most heavily Catholic areas, as in the more Protestant district of Limavady: there are too few occupations of a well-paid progressive nature altogether and, of what there is, Catholics get a smaller than proportionate share.

Only in teaching and nursing, and in the lower-grade service jobs, do they have an excess of proportions over their whole share of the jobs markets. The most that can be said about this is that they are found in occupations where personal services are rendered, very often along denominational lines, and where the services are more generally available to the public. Protestant managers are much in evidence everywhere, as are Catholic lower grades, even in Protestant areas.

Summary of Findings in Chapters 7 and 8

We have examined, in detail, the employment status and occupational orders in all 26 districts, a mass of printout which seems to consist largely of zeros or very small numbers (especially for women) at the district level. But this exercise has yielded little over and above what we already knew from the published national figures. There are, however, some local variations which could neither be predicted, nor all be explained in retrospect, from the total statistical information: different methods of labour market investigations would be needed to explain everything away.

However, some quite positive conclusions emerge. The first of these is that the less than proportionate representation of Catholics shows up most clearly in industries (and occupations) where large establishments account for a high proportion of all jobs. Where the employment units are small, whether in self-employment categories, or retailing, or craft production, Catholics form something more like the proportion of the employed population which we would expect from their overall share.

The second conclusion is that the figures always show a much better balance if we calculate the proportions in each category as a share of the total economically active and employed population. If we look at them as a share of the potential workforce (the over 16s), the picture is much less favourable for the Catholic minority. This is further accentuated if we take people in the relatively well-paid or more secure jobs as the proportion of the whole population. Thus from our earlier analyses of the Catholic age structure, of economic activity rates, of the proportion of older, prematurely retired people, and of course of the unemployment picture, the socio-economic structure of the Catholic minority becomes much more alarming: far fewer men and women within the Catholic population had, by 1981, secured positions of authority, with higher status and better pay, than would be expected when calculating their share of the adult population of Northern Ireland.

Thirdly, we find that disaggregation to local level does produce some interesting results. Mobility is clearly limited: labour markets must be even smaller than the official travel-to-work areas, and we shall examine in Chapter 10 some of the causes and implications of this low mobility.

Next, we come up against the problem of the age structure of the employed population. The figures given are, in theory at least, consistent with the view that the main changes which gave more opportunities to Catholics occurred only after 1965: thus positions of authority (professional and managerial, foremen and charge hands) would mostly be held by middle-aged or older workers. If certain industries and occupations only became open to Catholic entrants about that time, by 1981 they would still be under 35 years of age, and it could be that their lowly position on the ladder was in the process of changing at the time of the Census. This can only be investigated when we have the results of the 1991 Census. Neither the Labour Force Survey, nor the Continuous Household Survey, enables us to say for certain that such a movement is taking place, though the claim is often heard. In this context, the examination of the educational system in the next chapter should provide us with some clues.

Lastly, it is worth looking at the figures for the elucidation of the relationship between local denominational structure and the occupational configuration. In a predominantly Protestant area, we would expect more of the better opportunities to go to Protestants than to Catholics – whether through simple discrimination or by the complex recruitment mechanisms which form such a notable part of the Northern Ireland employment scene. This hypothesis is borne out by local figures, with the possible exception of the concentrations of high-status residents in the outer Belfast area generally, as in Castlereagh and North Down.

When we turn to the more heavily Catholic areas, we might expect to find a mirror image: that is, if there are more Catholics among the employers, local government officials, and other high-status groups, there could then be a corresponding higher proportion of middle-status groups for other Catholics (assuming either that they were given preferential treatment, or that normal recruitment procedures would automatically ensure their selection). This, however, does not turn out to be the case. Catholics are underrepresented in the better positions in all grades even in predominantly Catholic areas, just as they are much more likely to be unemployed. It appears to make little difference to the employment prospects of Catholics whether, at the local level, they are a very small minority or a much more substantial proportion of the population.

From this last observation it again follows that we have to look much more closely at the educational and training experience of the denominational groups, to see whether there may be other explanations for this persistent, and almost universal, differential in prospects.

9
Education, Training and Employment Chances

In this chapter we shall provide only a rough overview of the situation as it affected young people in Northern Ireland in 1981. The question of education in relation to employment has been extensively reviewed by Robert D. Osborne and his colleagues (Cormack and Osborne, 1983; Miller and Osborne, 1983; Osborne and Murray, 1978; Osborne et al., 1983; Osborne, 1985; Osborne and Cormack, 1986, 1987), and nothing that is said here should be taken as implying that their analysis is in any way mistaken.

We address here a fairly narrow question: what evidence is there, partly from the 1981 Census, partly from statistics issued by the Department of Education for Northern Ireland (DENI) on the basis of returns from the local Education and Library Boards, of differences in educational participation rates, in attainments, in the choice of subjects in further education, as between districts, and as between the denominations? How far could we explain differential employment prospects in terms of the available information on schools and colleges?

We have to start with a number of important caveats concerning the statistics, most of which have been fully spelled out in the work of Osborne and his colleagues (for example, the fact that the 1971 and 1981 Census data cannot be compared in a number of important respects). For the purposes of this chapter, the most important limitation encountered when relating education and employment to religion is the time lag between the childhood and adolescence of the workers enumerated in 1981, and their age at the Census. Thus the great majority of workers in occupations which might be dependent on qualifications would be between 25 and 54 in 1981. This means that the oldest of them would have left school in 1951, and if they did participate in further or higher education this would have occurred before the 1961 Census. Only the youngest cohort (25–29) would have completed their secondary education by 1971, and only part of the next older age group (30–34) would have completed degree-level education by the time of the 1971 Census. Thus the position in 1981 refers to an educational system which has been superseded, and for that reason the educational analysis for the Census year (and in some cases subsequent observations relating to school-leavers and educational participation rates) could only be useful in answering the question; what are the prospects, by religion, for those who were just finishing their education whilst this report was being written?

If we break down the percentage of persons with any sort of educational qualification by age, in the 1981 Census, we find that in the youngest group (16–25) 72 percent of Protestants were in possession of a certificate of some sort, and 61.9 percent of Catholics. But for all age groups the proportions were lower: 45.4 percent of Protestants, 39.4 percent of Catholics. So performance has improved, but the differential remains. It is also true that unemployment percentages rise as the level of qualification drops, so the legacy of past inadequacy will be with us for a long time (Osborne and Cormack, 1987: 82 ff., based on Continuous Household Survey).

Even this narrower approach suffers from a serious limitation when comparisons are being made: those who had entered the labour market between 1971 and 1981 did so at a time of relatively full employment, whilst their successors were faced with widespread unemployment, as has been shown.

If, therefore, there has been a narrowing of the gap between educational attainment rates, this would not necessarily be reflected in employment chances, and local differences would be more pronounced if better opportunities were confined to Greater Belfast (as was largely the case) and a high proportion of the improvement in educational participation rates and further education chances occurred outside that area. This will be seen to be specially relevant in the case of the Catholic population, and it increases the importance of personal mobility. if we are looking to improvements in educational provision helping to overcome the local unemployment differential, then the question of the likelihood of young people moving to areas of greater opportunity will turn on the physical and psychological obstacles to movement from one area to another.

Unfortunately much of the work by Osborne and his colleagues refers to national aggregates only, and has, since the 1981 Census, had to rely on the Continuous Household Survey and the annual Labour Force Survey, which do not usually allow a breakdown to district level, and only rarely to area groups. Some local studies have been undertaken (Cormack et al., 1980; Murray and Darby, 1980; Osborne, 1984, 1985), but a general analysis of the local picture is possible only on the basis of Census data.

Qualified Population

In Table 9.1 we look at the general structure of qualifications by religion, without regard to age. (For the most detailed national

Table 9.1 *Northern Ireland population aged 18 years and over by sex, religion and educational attainment, 1981*

	All denominations				Roman Catholic				Protestant and others				Not stated			
	Males		Females		Males		Females		Males		Females		Males		Females	
	no.	%	no.	%	no.	%	no.	%	no.	%	no.	%	no.	%	no.	%
Total population 18+	482,933		527,063		124,643		136,649		273,117		303,593		85,173		86,821	
Unqualified population	442,761		482,970		116,066		123,978		248,774		279,421		77,921		79,571	
Qualified population[1]	40,172	8.3	44,093	8.3	8,577	6.9	12,671	9.3	24,343	8.9	24,172	29.0	7,252	8.5	7,250	8.3
Subject group[2]																
Education	4,893	12.2	12,971	29.4	1,936	22.6	4,096	32.3	2,128	8.7	7,008	0.3	829	11.4	1,867	25.7
Health, medicine, dentistry	5,123	12.7	18,615	42.2	1,313	15.3	5,694	44.9	2,958	12.1	9,976	41.2	852	11.7	2,945	40.6
Technology and engineering	8,455	21.0	181	0.4	985	11.5	30	0.2	6,003	24.7	109	0.4	1,467	20.2	42	0.6
Agriculture, forestry, vet.	950	2.4	101	0.2	126	1.5	9	0.1	683	2.8	71	0.3	141	1.9	21	0.6
Science (inc. maths and applied)	4,465	11.1	1,885	4.3	843	9.8	399	3.1	2,649	10.9	1,145	4.7	973	13.4	341	4.7
Social, business and admin. studies	9,346	23.3	3,513	8.0	1,721	20.1	850	6.7	5,899	24.2	1,920	7.9	1,726	23.8	743	10.2
Vocational (inc. architecture and other prof. studies)	1,317	3.3	1,172	2.6	176	2.0	220	1.7	877	3.6	796	3.3	264	3.6	156	2.1
Language (lit. and area) studies	1,848	4.6	3,041	7.0	604	7.0	739	5.8	889	3.6	1,671	6.9	355	5.0	631	8.8
Arts (excl. lang. and perf. arts)	3,076	7.6	1,652	3.8	778	9.1	449	3.5	1,837	7.5	876	3.6	461	6.4	327	4.5
Music, drama and visual arts	699	1.7	962	2.2	95	1.1	185	1.5	420	1.7	600	2.5	184	2.5	177	2.4

[1] Percentage is of total population 18+.
[2] Percentage is of those qualified; percentages may not always add up to 100 due to rounding.

Source: Northern Ireland Census 1981, *Religion Report*, Table 10

overview see Osborne, 1985.) The term 'qualifications' is an imprecise one, and from necessity we follow here the groupings used in the Census (Northern Ireland Census 1981, *Religion Report*: xii). That is to say, we include in the qualified population only those who have completed courses dependent on the possession of GCE A-level passes, whether degrees, diplomas or certificates. This definition excludes many qualifications below this highest level, for example diplomas and certificates which may be dependent on, or equivalent to, GCE O-level passes, and this involves many manual and non-manual occupations which are normally classified as 'skilled' (socio-economic groups III(N) and III(M)). However, as a first approximation this classification will suffice if we are merely looking at the top echelons of the labour market – under 10 percent of the total population aged 18 years and over.

The Continuous Household Survey uses a much broader classification of 'qualifications', and shows over 40 percent of the population as possessing some evidence of educational attainment (Osborne and Cormack, 1987: 83). The table appears to show that a smaller proportion of Catholic males had any kind of qualification in 1981, than others, but a higher proportion of females, and this confirms the picture presented in Chapters 7 and 8. Here we have another case where the 'not stateds' are nearer to the whole population than to the Catholics in their composition, making it clear once more that we cannot apportion the 'not stateds' to denominations on a different basis for each parameter, by looking at which religion group bears the strongest resemblance to the group not responding to this question.

Going down the list of types of qualifications, it is immediately apparent again that Catholics have a disproportionately high share of occupations connected with education and health, and it is the 45 percent of qualified Catholic women in health – that is, the nurses – who account for the high overall total. As soon as we turn to the technical subjects (including engineering, science, business studies) the Catholic population appears to be less well represented, and only in language and literature studies and the other traditional arts and humanities subjects does the proportion of qualified Catholics rise. The total numbers involved in these categories are small. Over 90 percent of the adult population was unqualified in the Census definition of the term. Non-response or error would not alter the picture. Perhaps 1,500 more Catholic men would have been available for jobs requiring technical or professional qualifications, had their acquisition of such skills been available to them in the previous decades. This might have conceivably had an effect on their employment status distribution, which we have already discussed.

Let us now see (Table 9.2) how this national picture is reflected in

Table 9.2 *Northern Ireland population aged 18 years and over by sex, religion and educational attainment, in area groups, 1981 (percentages)*

Area groups[1] and attainment	All denominations		Roman Catholic		All other stated denominations		Not stated	
	Male	Female	Male	Female	Male	Female	Male	Female
I Belfast								
Total qualified	10.6	8.5	9.3	9.5	10.6	8.0	11.7	9.6
Prof. qual.	6.4	6.4	5.4	7.3	6.4	6.1	7.0	6.5
Soc. bus. qual.	2.8	0.8	2.4	0.8	2.9	0.7	3.1	1.2
Lang., arts etc. qual.	1.4	1.3	1.6	1.4	1.3	1.2	1.6	1.8
Population 18+	219,468	252,192	33,658	40,221	147,397	171,424	38,413	40,547
II Northern								
Total qualified	7.2	8.7	6.4	10.7	7.5	8.1	7.0	8.4
Prof. qual.	4.8	7.2	4.3	9.2	4.9	6.7	4.7	6.7
Soc. bus. qual.	1.4	0.6	1.1	0.6	1.5	0.5	1.3	0.6
Lang., arts etc. qual.	1.1	0.9	1.0	0.9	1.1	0.9	1.0	1.0
Population 18+	78,643	83,627	16,503	17,682	49,689	53,296	12,451	12,649
III Southern								
Total qualified	6.3	8.4	6.1	9.6	6.7	7.8	5.6	7.1
Prof. qual.	4.1	7.0	4.0	8.1	4.5	6.5	3.7	5.7
Soc. bus. qual.	1.2	0.6	1.0	0.6	1.3	0.5	1.1	0.5
Lang., arts etc. qual.	1.0	0.9	1.1	0.9	0.9	0.8	0.8	0.8
Population 18+	109,345	115,279	40,202	42,891	48,369	51,710	20,774	20,678

Table 9.2 *continued*

Area groups[1] and attainment		All denominations		Roman Catholic		All other stated denominations		Not stated	
		Male	Female	Male	Female	Male	Female	Male	Female
IV Western									
Total qualified		5.8	7.7	5.6	7.8	6.3	8.0	5.3	6.6
Prof. qual.		3.9	6.4	3.6	6.7	4.3	6.6	3.5	5.3
Soc. bus. qual.		1.0	0.4	1.0	0.4	1.1	0.5	1.1	0.5
Lang., arts etc. qual.		0.9	0.8	1.0	0.7	1.0	0.9	0.6	0.8
Population 18+		75,477	75,965	34,280	35,855	27,662	27,163	13,535	12,947
Northern Ireland									
Total qualified	(no.)	40,172	44,093	8,577	12,671	24,343	24,172	7,252	7,250
	(%)	8.3	8.4	6.9	9.3	8.9	8.0	8.5	8.4
Prof qual.	(no.)	25,203	34,925	5,379	10,448	15,298	19,105	4,526	5,372
	(%)	5.2	6.6	4.3	7.6	5.6	6.3	5.3	6.2
Soc. bus. qual.	(no.)	9,346	3,513	1,721	850	5,899	1,920	1,726	743
	(%)	1.9	0.7	1.4	0.6	2.2	0.6	2.0	0.9
Lang., arts etc. qual.	(no.)	5,623	5,655	1,477	1,373	3,146	3,147	1,000	1,135
	(%)	1.2	1.1	1.2	1.0	1.2	1.0	1.2	1.3
Population 18+		482,933	527,063	124,643	136,649	273,117	303,593	85,173	86,821

[1] For definition of area groups see Appendix A.

Source: Own calculations based on Northern Ireland Census 1981, *Religion Report*, Table 10 (unpublished data) 'Population 18+ and over by religion and educational attainment'

local areas. We have here reduced the categories by dividing those qualified into only three groups: the professions (which include most of the technical qualifications); social science and business qualifications; and arts or humanities. Belfast stands out, as we would expect, as having the highest proportion of qualified men, but the same is not true of women. Belfast's population stands out as being better qualified than is the case elsewhere. For Catholic men this difference is even more pronounced: 9.3 percent of them fall into the 'qualified' category, about 50 percent more than the average for the other area groups. Other denominations are very close to the average, and the 'not stateds' are more like the Protestants and others than like the Catholics. We do not give the numbers here which are represented by differences in percentages, but for each gender, area group and qualification group they are quite small, and the differences between denominations smaller still; this is true even of Belfast, but *a fortiori* of the other area groups. As soon as we are outside Greater Belfast, the percentage of qualified women exceeds that of qualified men; and this is almost entirely due, once again, to teaching and nursing. The other categories affect an almost negligible proportion of the total labour force, and local variations are insignificant; without education and health, almost the whole Catholic population would lack qualifications. Given the denominational links of these two sectors, we do not expect representation to be greatly out of line.

School-Leavers' Qualifications

In Tables 9.3 and 9.4 we show recent statistics comparing qualifications below those shown in the earlier tables, both over time, and by Education and Library Board areas for 1981–82 (see also Figure 22). These tables are not broken down by religious affiliation, but, given the distribution of Roman Catholics as between the area groups, they throw some light on the local educational picture up to the early 1980s. These figures in themselves cannot be used to predict local employment chances; these will depend on the relevance of the qualifications attained to the job opportunities offered. Nevertheless it is true in general that unqualified leavers, or those with poor qualifications, will find it more difficult to get jobs; they will have lower job mobility, at any rate within Northern Ireland, and in particular they will have less access to the expanding sectors which are heavily concentrated on Greater Belfast.

Table 9.2 shows time trends for Northern Ireland. One notes that these figures, and trends, compare well with Great Britain (Central Statistical Office, 1986: Table 5.5). Table 9.2 is directly comparable with a similar table prepared for the rest of the UK. Time trends are

Table 9.3 *School-leavers' qualifications, Northern Ireland, 1975–76 to 1981–82*[1]

	1975–76	1976–77	1977–78[2]	1979–80[2]	1981–82[2]
Boys					
2 or more A levels	2,009	2,141	2,244	2,120	2,311
1 A level	464	504	410	470	494
5 or more O levels[3]	1,005	1,143	1,079	1,216	1,253
1–4 O levels[3]	2,645	2,818	2,324	2,374	2,763
Low grades[4]	—	—	—	—	2,900
No qualifications[5]	7,330	7,457	7,673	7,927	4,809
Girls					
2 or more A levels	1,972	2,016	2,142	2,265	2,469
1 A level	490	537	458	468	590
5 or more O levels[3]	1,137	1,282	1,203	1,643	1,524
1–4 O levels[3]	2,982	3,310	2,774	2,788	2,887
Low grades[4]	—	—	—	—	2,753
No qualifications[5]	6,367	6,396	6,256	6,270	3,000
All school-leavers					
2 or more A levels	3,981	4,157	4,386	4,385	4,780
1 A level	954	1,041	868	938	1,084
5 or more O levels[3]	2,142	2,425	2,282	2,859	2,777
1–4 O levels[3]	5,627	6,128	5,098	5,162	5,650
Low grades[4]	—	—	—	—	5,653
No qualifications[5]	13,697	13,853	13,929	14,197	7,809

[1] Includes an element of school-leavers transferring to institutions of further education to complete their full-time education.
[2] Figures exclude leavers from GCE courses at institutions of further education.
[3] Grades A–C only and grade 1 CSE.
[4] CSE grades 2–5 and GCE O-level grades D–E.
[5] Includes those leavers with CSE grades 2–5 and those with grades D–E at GCE O level prior to 1981–82. From 1981–82 includes only those who undertook no GCE/CSE examinations or obtained no graded results.

Source: Northern Ireland *Annual Abstract of Statistics* no. 3, Table 5.6

the same, but Northern Ireland stands out as having a better record, in regard to educational qualifications, than the rest of the UK. Whether this is because more money has been spent, per caput, on education than on other fields of government expenditure; whether there is a fundamentally different attitude to education, perhaps in part encouraged by the churches; or whether there are long-term difficulties in finding jobs in Northern Ireland, or in countries to which the Irish migrated, we do not know. At any rate these figures matter because they dispose of the argument that the people of Northern Ireland, as a whole, are underqualified. It can in fact be shown that this applies to the Catholics, taken on their own: their educational attainments are better than those observed at least in

Table 9.4 *School-leavers' qualifications by Education and Library Board divisions, 1981–82 (numbers and as a percentage of all school-leavers in Board)*

Qualifications[1]	Belfast no.	Belfast %	Western no.	Western %	North Eastern no.	North Eastern %	South Eastern no.	South Eastern %	Southern no.	Southern %
3 or more A levels	999	15.6	427	8.9	625	10.6	523	11.3	574	12.6
1 or 2 A levels	680	10.6	478	10.0	579	9.8	433	9.4	546	12.0
Higher-grade O levels[2]	1,685	26.3	1,543	32.2	2,107	35.6	1,600	34.6	1,482	32.5
Lower-grade O levels and CSE[3]	1,341	20.9	1,015	21.2	1,172	19.8	1,246	26.9	869	19.0
No qualifications[4]	1,712	. 26.7	1,324	27.7	1,437	24.3	826	17.8	1,093	23.9
Total	6,417		4,787		5,920		4,628		4,564	

[1] Special and independent schools excluded. Fifteen year olds transferring to further education institutions to complete their compulsory education included. Percentages may not always add up to 100 due to rounding.
[2] O-level grades A–C and CSE grade 1.
[3] O-level grades D–E and CSE grades 2–5.
[4] Ungraded O level and CSE and those who undertook no GCE/CSE examinations.

Source: Own calculations based on unpublished tables of school-leavers, Department of Education, Northern Ireland

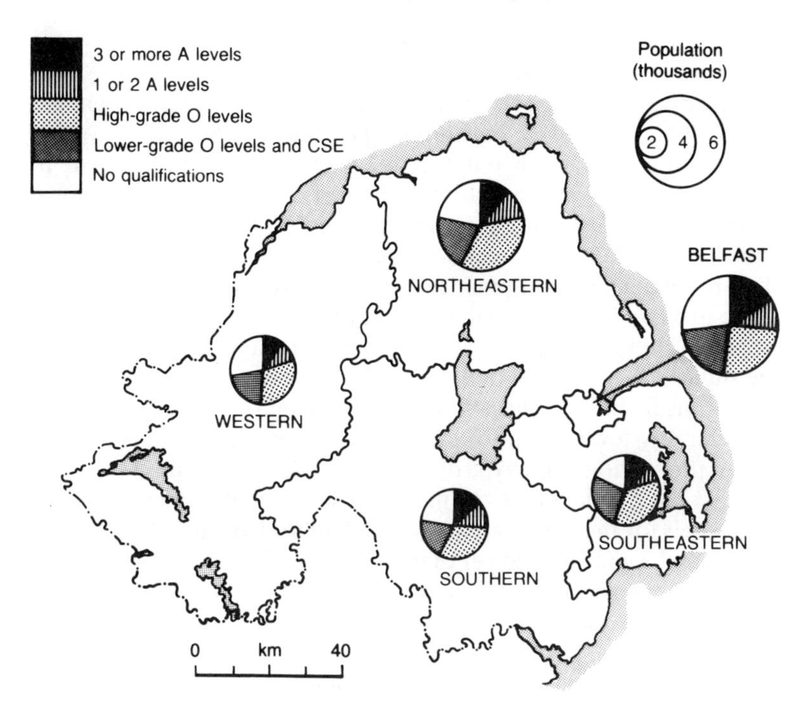

Figure 22 *School-leavers' qualifications by Education and Library Board divisions, 1981–82*

Source: Own calculations based on unpublished tables of school-leavers, Department of Education, Northern Ireland

England and Wales. (Scotland again has a better educational record.) The question remains whether this overall propensity to participate in education is in itself related to employment levels. Certainly, education has grown whilst employment has declined.

The evolution over time is clear. Between 1975–76 and 1981–82 the proportion of those who left school with at least one A level has risen from 18.7 percent to 21.1 percent, and that compares with 18.1 percent in the UK overall (in 1981–82). In 1975–76 a further 8.1 percent had five or more O levels, and in 1981–82 the percentage was 10. In the UK this was fractionally lower. The lower O level and CSE ratings are not available for 1975–76, but equipped a further 40.7 percent of the school-leavers, leaving about 30 percent with no qualifications at all. In the UK as a whole that proportion was indeed a good deal lower, at 12.8 percent, but there has been a change in counting 'other' grades and awards, and on pre-1980 counts (where anything below one to four O levels and CSE grade I was lumped together with the 'no qualifications'), the figures are remarkably similar in the two areas. At the top end of the qualifiers' table Northern Ireland does remarkably well, but it is likely that there were also relatively more leavers at the bottom of the distribution with no pieces of paper at all, or only nominal certificates.

How does this work out locally? The DENI figures relate to different areas from those used in the rest of this report. (The local figures are taken from a different source than those for the whole country, and there is a slight discrepancy in the totals of leavers, but the proportions remain practically unaffected. The explanation lies in the first two footnotes of the DENI table for the whole of Northern Ireland.) Belfast is Belfast city only. South Eastern is dominated by the suburban districts on that side – North Down, Ards, Castlereagh – and unsurprisingly this shows up in our table with at any rate a considerably lower proportion of totally unqualified leavers, and a higher proportion of the type of leaver we would associate with future white-collar employment. At first sight surprisingly, the percentage of A levels for the Board area is lower than for Northern Ireland as a whole, and a good deal lower than for Belfast, but this is entirely due to the presence in Belfast of a large number of public sector grammar schools whose catchment area includes the adjoining suburbs.

The North Eastern area also includes some Belfast suburbs, but they are not as favourable in terms of school-leaving qualifications as the South Eastern ones, a distinction we have previously seen when looking at district socio-economic data. The Southern area (which is more heavily Catholic than the two Eastern areas) comes next; then the Western area, which is still more predominantly Catholic, does least well, with the lowest proportion of leavers with A levels and the highest of unqualified leavers. It is not, however, notably different from the rest of the country as regards proportions in the leavers who could expect to get the better manual and the white-collar jobs, where the percentages are equal to the national average and exceed those for Belfast. So, on these figures alone, the educational standards of the West would seem to leave room for improvement, taking school-leavers alone, and disregarding participation in further education, where, it appears, more pupils in the Western Board area continued their studies.

The structure of education in Northern Ireland is very much more complex than in England, with denomination as the deciding factor. Almost all Catholic children are in maintained voluntary schools at the primary and the secondary (intermediate) stage, and in voluntary grammar schools under Catholic management later. Protestant children are in controlled schools at the primary and secondary (intermediate) level, and in controlled or voluntary grammar schools which, though not explicitly governed by Protestant bodies, do in fact have an overwhelming number of Protestant governors. (This applies whether or not the name of the school denotes a denominational affiliation.) There are some children from Catholic schools in these effectively Protestant schools; and by the mid 1980s there were the beginnings of a movement towards interdenominational secondary education, but its significance was small. There are one or two notable exceptions: for example in Limavady, we were told, the maintained grammar school has 30 percent Catholic students; there is no Catholic grammar school there, and those local Catholic children who do not go to the maintained school travel to Coleraine rather than Derry. Thus figures at the most local level may be misleading, and it would be best to confine one's interpretation to the area group figures. Even these are not self-contained, as the case of Limavady residents at a Coleraine school shows, and for the Greater Belfast area the district figures are even less reliable. By and large, we can assume that children in Catholic-controlled schools are Catholics, and that children in 'other' schools belong to 'other' denominations (Osborne, 1985: 20).

Table 9.5 *School pupils in Northern Ireland by type of school, by Education and Library Board divisions, January 1982*

Education and Library Board division	Controlled: Secondary (intermediate)	Controlled: Grammar (excluding prep.)	Maintained voluntary: Secondary (intermediate)	Maintained voluntary: Grammar (excluding prep.)	Total
Belfast	9,567	2,093	11,838	12,458	35,956
Western	5,568	2,300	12,466	7,221	27,555
North Eastern	15,702	4,041	8,297	8,615	36,655
South Eastern	13,742	2,801	5,373	5,991	27,907
Southern	9,442	2,437	13,103	6,521	31,503
Total	54,021	13,672	51,077	40,806	159,576

Source: Own calculations based on unpublished tables from the School Census, Department of Education, Northern Ireland

In Table 9.5 the first two columns therefore show only Protestant children, the third column only Catholic children, and the fourth one is mixed (though segregated). Despite recent changes, one still hears the charge that the voluntary grammar schools have been slower to adapt to new conditions than the controlled schools. But in Belfast, where school-leavers' qualifications are particularly high, the majority of children are in such voluntary schools. So it may be more a matter of the socio-economic strata from which the schools draw their pupils than their religious affiliation. In the North Eastern and South Eastern areas, which are dominated by the mainly Protestant Belfast suburban areas, and the Protestant rural north generally, controlled secondary schools are more important. In the Southern and Western areas (which are more predominantly Catholic) the voluntary sector predominates as much as in Belfast. Thus, detailed examination of the record of various schools apart, there is nothing in this structure to explain the variations in examination results shown above.

Further Education

In Tables 9.6, 9.7 and 9.8 we examine the structure of further education in the Province. (For a detailed investigation of the further education system in Northern Ireland, see House of Commons, Education, Science and Arts Committee, Session 1982–83, *Further and Higher Education in Northern Ireland*, two volumes (Report and Minutes of Evidence): HMSO, 1983, HC 180–I and HC 180–II.)

Table 9.6 *Students in vocational courses of further education in area groups, advanced and non-advanced, full-time and part-time, 1981–82*

	Advanced		Non-advanced	
Area group[1]	Full-time	Part-time	Full-time	Part-time
I Belfast	188	733	3,498	15,667
II Northern	29	64	2,450	2,981
III Southern	—	51	4,655	4,353
IV Western	105	223	2,672	4,133
Northern Ireland	322	1,233[2]	14,012[2]	27,506[2]

[1] For definition of area groups see Appendix A.
[2] These totals include some institutions not allocated to area groups, but exclude the Ulster Polytechnic.

Source: Own calculations based on unpublished tables of further education statistics, Department of Education, Northern Ireland

In Table 9.6 we have used our four area groups to make comparisons, because it was possible to allocate most institutions to a particular district (though the boundaries of catchment areas are obviously not so precise). This shows, once again, the concentration of effort on Belfast. Of 322 full-time advanced students in vocational courses, 58.4 percent were in Greater Belfast, nearly 60 percent of the part-time advanced students. In non-advanced vocational education, the contrast is less striking: Belfast accounts for only 25 percent of the full-time courses, and under 60 percent of part-time courses.

Nevertheless, in relation to the size of each age group, as we shall see, the imbalance is probably no more than we should expect when comparing a metropolitan with a non-metropolitan area.

In Table 9.7 we look at the national structure of further education by subject, and in relation to the relevant age group. These figures should be read in conjunction with the earlier ones on school-leaving qualifications. It is, of course, impossible to say what proportion of those who left school without paper qualifications in 1981–82 will, sometime in the future, form part of the further educational analysis. Only for the advanced group may it be safely assumed that participants either did non-advanced work previously, or had at least O-level school-leaving successes.

Total numbers in advanced courses are clearly very small, and the table shows that whereas full-time students are heavily concentrated on business and social studies (a high proportion, in fact, in accountancy, and in social work qualifying courses), part-timers are more likely to be in technological studies (which include computer studies in most cases). The proportions involved in these types of courses, as a part of the age group 15–24, are negligible; to some extent they form a complement to the (much larger) university and polytechnic group, which does not form part of this report at all, and we cannot really include them in our study of the labour market. This is not to say that higher education does not play a role in providing advantages within Northern Ireland, or indeed the UK labour market, but questions of local variations, or religious denominational representation, do not arise in the same way. These matters have been fully dealt with elsewhere (McCartney and Whyte, 1984; Northern Ireland Economic Council, 1985; Cormack et al., 1984).

We therefore concentrate on the non-advanced sector, in which 14.7 percent of all 15–24 year olds were involved in 1981–82. Since we may, as a first approximation, assume the duration of each course in the non-advanced sector to be between one and two years (without allowing for whether these are full-time or part-time courses), then this percentage (of current students against a ten-year age group) is clearly quite considerable. If all 40,000 enrolled students were 17 and 18 years old, then this would mean that two-thirds of them were doing some kind of course. This is not so in practice, but the figure of 40,000 in any one year is large enough to allow the speculation that well over half the 'early' school-leavers get some kind of further education, though not necessarily a paper qualification. The true total is probably much higher. If, in fact, we add together the 40,000 students in institutes of further education, the 6,600 in the Ulster Polytechnic, the 1,400 in teacher training colleges outside universities and polytechnics, and the 8,000 university students, a total educational parti-

Table 9.7 *Students in advanced and non-advanced further education courses, Northern Ireland, 1981–82 (numbers and as a percentage of population aged 15–24 years)*

	Advanced				Non-advanced				
Course	Full-time	Part-time	Total no.	Total %	Full-time	Part-time	Total no.	Total %	Total %
Technology	33	586	619	0[1]	2,336	9,131	11,467	4.2	4.4
Bus. admin. social	161	359	520	0	4,221	6,873	11,094	4.1	4.2
Other	128	126	254	0	6,718	11,130	17,848	6.5	6.6
Total	322	1,071	1,393	0.51	13,275	27,134	40,409	14.7	15.2
Building	—	169	169	0	615	2,063	2,678	1.0	1.0
Other bus. commerce	126	148	274	0	1,981	1,704	3,685	1.4	1.5
Secretarial	35	36	71	0	2,049	3,939	5,988	2.2	2.2
Acc. bank insurance	—	—	—	0	—	859	859	0	0
Total[2]	161	353	514	0	4,645	8,565	13,210	4.8	5.0

[1] Zero indicates nil or negligible.
[2] The total here is included in the first total.

Source: Own calculations based on Department of Education, Northern Ireland, unpublished computer printout

cipation rate is reached for Northern Ireland which compares very well with UK averages (Central Statistical Office, 1986: Table 5.6).

The question then still is: where are they, who benefits and how relevant is what is taught? We may exclude from this question polytechnics, teacher training institutions and universities, partly because their potential labour markets are so much wider, and partly because at least the polytechnic and teacher training element is centrally managed and therefore presumably sensitive to overall demand, taking a longer-term view. In what follows we concentrate on the further education sector because that will be the one which will play the most crucial role in adjusting labour supply to local labour markets.

In Table 9.8 we show the area breakdown of further education in detail (once again allocating each institution to the district in which it is situated, remembering that in the Greater Belfast area in particular there will be considerable travel to education from outside the immediate suburbs which are included in area group I). We exclude students in advanced courses, because the national total (under 1,400 students) is too small to produce a worthwhile breakdown by subject area.

Looking at the totals in each area group, as a proportion of all 15–24 year olds, Belfast stands out with 16.5 percent of the age group in further education of all kinds (Figure 23). Second, however, comes the Western area, which is geographically truly distinct; it has 15.3 percent of the age group in further education, with relatively the same proportions as Belfast in advanced courses. The Southern group comes a little way behind, and the Northern one is a good deal lower still, with only 12.9 percent in the further education sector. Whether that is because of considerable travelling from districts like Antrim, Ballymena and Larne into Belfast area colleges, or whether it is a real difference, we cannot say.

One striking difference, however, is immediately apparent. In Belfast the overall high figures are almost solely due to the large numbers of part-time students who form over 80 percent of all students. In the South there are more full-time than part-time non-advanced students, and in the North only a few less in full-time than in part-time education; in the West 60 percent are full-time students. This would appear to indicate that part-time studies (that is, day release and evening courses) are a predominantly urban phenomenon, and particularly so in Belfast with its high proportion of public sector, banking, financial and other commercial institutions, where day release is normal and evening courses are more feasible than in less densely populated areas. Closer inspection of the figures, however, shows that the financial sector does not by any means account for all the Belfast excess in this category, with only about a quarter of the part-time students coming from the most obvious source. In fact the detailed breakdown returns of further education participants shows a large number in a variety of service industries: retailing, hairdressing, catering and so on. In Table 9.8 we have picked out a few subsectors, namely building, other business etc., secretarial, and accountancy etc., with the totals for these included in the overall

Table 9.8 *Students in non-advanced further education courses, Northern Ireland, by area groups, by main subject groupings, 1981–82 (numbers and as a percentage of the area population aged 15–24 years)*

Area group[1]	Full-time	Part-time	Total no.	Total %	Total[2] %
I Belfast					
Technology	697	5,534	6,231	5.1	5.5
Bus. admin. social	1,056	3,761	4,817	3.9	4.2
Other	1,745	6,372	8,117	6.6	6.8
Total	3,498	15,667	19,165	15.7	16.5
Building	42	801	843	0.7	0.8
Other bus. commerce	417	1,055	1,472	1.2	1.3
Secretarial	526	1,512	2,038	1.7	1.7
Acc. bank insurance	—	744	744	0.6	0.6
Total[3]	985	4,112	5,097	4.2	4.4
II Northern					
Technology	536	1,018	1,554	3.6	3.6
Bus. admin. social	961	914	1,875	4.4	4.6
Other	953	1,049	2,002	4.6	4.6
Total	2,450	2,981	5,431	12.6	12.9
Building	111	254	365	0.9	0.9
Other bus. commerce	496	184	680	1.6	1.8
Secretarial	359	655	1,014	2.4	2.4
Acc. bank insurance	—	55	55	0[4]	0
Total[3]	966	1,148	2,114	4.9	5.2
III Southern					
Technology	665	1,018	1,683	2.6	2.7
Bus. admin. social	1,190	1,276	2,466	3.9	3.9
Other	2,800	2,059	4,859	7.6	7.6
Total	4,655	4,353	9,008	14.2	14.3
Building	307	440	747	1.2	1.2
Other bus. commerce	578	257	835	1.3	1.3
Secretarial	722	1,169	1,891	3.0	3.0
Acc. bank insurance	—	47	47	0	0
Total[3]	1,607	1,913	3,520	5.5	5.6

Table 9.8 *continued*

Area group[1]	Full-time	Part-time	Total no.	Total %	Total[2] %
IV Western					
Technology	438	1,561	1,999	4.3	4.5
Bus. admin. social	1,014	922	1,936	4.2	4.5
Other	1,220	1,650	2,870	6.2	6.3
Total	2,672	4,133	6,805	14.6	15.3
Building	155	568	723	1.5	1.6
Other bus. commerce	490	208	698	1.5	1.7
Secretarial	442	603	1,045	2.2	2.3
Acc. bank insurance	—	13	13	0	0
Total[3]	1,087	1,392	2,479	5.3	5.7

[1] For definition of area groups see Appendix A.
[2] This total includes the small number of advanced students in each area and subject.
[3] The second total in each area group is included in the first total.
[4] Zero indicates nil or negligible.

Source: Own calculations based on Department of Education, Northern Ireland, unpublished computer printout

totals above. It then turns out that although the Belfast dominance is quite apparent in some of the conspicuous occupations (744 out of the 859 national student body in part-time accountancy, banking and insurance courses are in Belfast), the dominance of the headquarters sector is not as strong as we might have thought, within the Belfast region, whereas the general and personal service sector, including building technology, are all very important.

Outside Belfast, no clear picture emerges. There are local differences, but they are not large enough to enable one to make policy-related statements. In the high-unemployment rural West too few young people are involved, perhaps, in the career-oriented courses, and practically none in the potentially high-earning financial sector.

We have looked at, but not analysed here, the range of courses offered in all institutions of further education (DENI, 1982). Clearly there are differences, dependent partly on size and partly on location. But on the whole it strikes one that a very wide range was offered in almost every college, and that the numbers involved in some of the courses (in 1981–82 returns) are so small that, in most of the rest of the UK, they probably would not be kept going at all. That much is evident even from the omnibus categories in the summary.

It is therefore not possible, on this kind of evidence, to claim that by 1981–82 the structure of further education in Northern Ireland was prejudicial to future employment prospects as such. We do not know, from the analysis, what proportions of people on each course

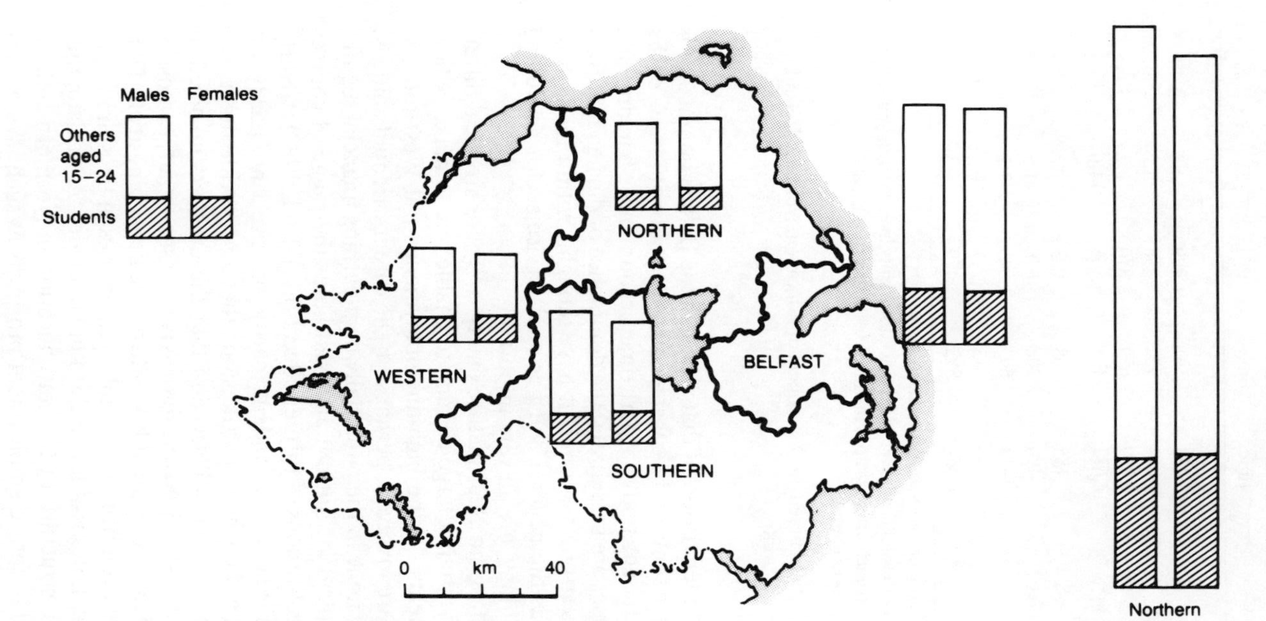

Figure 23 *Student population of Northern Ireland as a proportion of the population aged 15–24 years, by area group, 1981*

Source: Own calculations based on Northern Ireland Census 1981, *Economic Activity Report*, Table 4

were Catholics and others; however, since some of the colleges we looked at were located in mainly Catholic areas, it seems reasonable to suppose that access not only was theoretically possible but occurred in practice. Whether students always chose wisely, or whether the levels they reached were always sufficient, we do not know. What we do know, however, from our analysis of the industrial structure of the Western and Southern regions, and the more rural districts, is that the job market would not offer sufficient opportunities even for those who were being educated (except, possibly, in a few cases like the more highly trained secretaries, computer operators and book-keepers) without them travelling a very long way to work, or indeed moving to another part of the Province, or to Britain.

Lastly, we look at overall educational participation rates for full-time students only, by district, in relation to the whole population, by gender, aged 15–24 years (Table 9.9). This, it must be remembered, includes students at universities and polytechnics in Northern Ireland and some who were studying elsewhere in the UK, those at teacher training colleges, and those enrolled in institutions of further education, whether full-time, or effectively part-time but not otherwise employed. It should therefore be a fairly comprehensive category.

Table 9.9 *Students in population of Northern Ireland 1981, aged 15–24 years, numbers and as a proportion of the population, males and females, by area groups*

Area group[1]	Males 15–24	Male students no.	Male students %	Females 15–24	Female students no.	Female students %
I Belfast	58,094	13,848	23.8	56,950	13,306	23.4
II Northern	21,791	4,704	21.6	20,244	5,158	25.5
III Southern	31,756	7,149	22.5	29,041	7,630	26.3
IV Western	22,982	4,848	21.1	20,878	5,483	26.3
Northern Ireland	134,623	30,549	22.7	127,113	31,577	24.8

[1] For definition of area groups see Appendix A.

Source: Own calculations based on Northern Ireland Census 1981, economic activity tables

Table 9.9 is not broken down by religion. We have previously commented (see Tables 5.1 and 5.2) on the high overall proportion of Roman Catholics of all persons aged 16 years and over in education.

We see that 22.7 percent of all males of the ten-year age group and 24.8 percent of the women were involved in education. Again, on the assumption that the average length of each course (whether sixth form at school or anywhere else) was between one and two years, we would get a very respectable total participation rate. In Great Britain the corresponding male rate in 1981 was 20.9 percent, for women 20.8 percent, so Northern Ireland did well in this respect. (There are, in fact, a large number of other ways of measuring education participation rates – calculating full-time equivalents, giving single-year age- and sex-specific groups, and so on – but the outcome would be much the same.)

Within this total rather favourable picture there are local differences, as we saw in our previous analysis of economic status. (The districts mentioned individually do not figure in the area group summaries printed in Table 9.9; the information is contained in Additional Table A7.2.) Inevitably, it seems, North Down and Castlereagh lead the field. However, this time we must add Coleraine, which has a university campus; given the confusion over 'place of usual residence', a large number of students returned themselves as permanent Coleraine residents. That would cause some marginal distortion for both males and females. For women, however, there are a number of other districts where a very high proportion of the age groups were students: Armagh, Dungannon, Fermanagh, Magherafelt, Newry and Mourne, and Omagh. In fact all these are urban centres with a large rural catchment area, each of them with a technical college or college of further education, and some of them with a Catholic grammar school which would have a number of boarders who ought not to be, but may have been, returned as being usually resident. This factor would apply more to girls than to boys. But once again not too much attention should be paid to this degree of localization.

Very few districts show significant differentials against regional or national participation rates. If we have, apparently, differences in age participation rates between two adjoining districts such as Derry and Limavady, this can be due to the fact that one or two hundred students who, for convenience, lived in an adjoining district to receive their post-16 education at the time of the Census, can make 2 or 4 percent difference in the local rates. So all told the picture is relatively uniform, and it is certainly not possible, on the basis of this table, to make very sweeping assertions. In Table 9.9 we group the districts in the four area groups and this shows how small the differences are, except for a slight excess for men in the Belfast group, and a slight excess for women in the Southern and Western groups.

Summary of the Link of Education to Employment

The history of education in Northern Ireland, in relation to the structure of the labour market, has been amply documented. Until the data from the 1981 Census, and the related statistics from the Department of Education for Northern Ireland, the Continuous Household Survey and the Labour Force Survey, became available, the picture was clear. Catholics and Protestants had very different occupational structures at least until 1971. Longitudinal studies suggested that there was little upward social mobility for working-class Catholics, so that the changing composition of employment (especially in the public sector) made little impact on Catholic class structure (Cormack and Osborne, 1983: 222). Changes either in educational provision, or in the composition of the demand for labour, were not thought to be capable of altering this structure substantially.

The difference between men and women is also of long standing. As far back as records of religion and employment go, the opportunities for Catholic women in nursing and teaching meant that there were at least two recognized avenues for girls with good educational qualifications to obtain posts which commanded a relatively good salary (compared, say, with junior clerical posts), some prospect of promotion, and security. No such avenues were open to men.

The next permanent feature of the educational scene, it appears, is the rigid division by denominations, which continues to the present in spite of a number of attempts over the years to lessen segregation, at least in secondary education. Curriculum changes for individual schools have been in the hands of governing bodies which were not always aware of the wider implications of subject choices within the secondary schools system. There was, and is, a considerable burden on the further education system to ensure that workforce entrants' qualifications were such as to maximize their career opportunities.

In general fewer Catholics passed through the grammar school system than Protestants (Osborne, 1985: 22). This imbalance has been made good by Catholic transfers to non-Catholic grammar schools only to a very limited extent, in one or two localities. The fact that many grammar schools are fee-paying institutions, and may not be in the locality where the child's parents live, may be a deterrent to such transfers, as well as a more general reluctance to allow children to leave the Catholic system.

The difference in emphasis on science and mathematics (in Protestant schools) and on the humanities (in Catholic schools) is of long standing. Recent investigations show that the gap has narrowed, but it still exists (Osborne, 1985: 25).

It would be a great mistake, however, to believe that these differences, by themselves, account for the differentials in opportunities for work. We have already seen that the greatest increases in jobs occurred in the service sectors, and very few of those, whether at the top (professional/managerial) or the bottom end of the distribution (unskilled and semiskilled workers), require O- or A-level passes in mathematics or science. All the better-paid jobs in commerce and administration require a good measure of literacy, and the analyses performed on the relevant pass rates in the two denominational groups show no appreciable difference overall, with the Catholics even showing a marginally higher success rate in commercial subjects.

It might be claimed that if the Catholic schools produced more scientists and mathematicians, school-leavers might stand a better chance of obtaining work outside Northern Ireland. Nobody has seriously suggested, however, that education should be determined by the need to prepare pupils for foreign labour markets.

The fourth aspect of the interaction of education and employment is the geographical angle. It has been claimed that the failure to prepare children in the areas of higher unemployment amongst Catholics (in all peripheral regions, but especially in the South and West) accounts for the failure of school-leavers to obtain jobs. That this is not correct has again been shown by Osborne (1985: 33 ff.). Indeed, given the structure of the local labour markets in those areas, any greater emphasis on science and technology would again mean a kind of compulsion for the school-leaver to find work away from his or her home area – a demand which has never been articulated and would find no favour in any part of the world, with educators, parents or children. Whether such a requirement can be imposed by politicians cannot be discussed here. In the rest of the UK this problem began to come to the forefront of debate in the 1980s, but it has not been an issue in Northern Ireland. Catholic schools in the peripheral areas do as well as Protestant schools in preparing children for life anywhere, as regards basic numeracy and literacy, and it has never been postulated that schools should adapt their curriculum to purely local demand for labour. Thus a measure of migration has always been built into the system, and has been reflected in the past in net migration losses to the rest of the UK and overseas. This has been what we have called earlier the 'labour market failure'; it is not a failure on the part of the schools.

The next point concerns the destination of school-leavers. In recent years, the proportion going on to colleges of education, the polytechnics and the universities has grown. This growing percentage of school-leavers has, by definition, shown willingness to transfer to other areas to continue their education.

At the tertiary level, at least in the polytechnic and university sector, the opportunities for Catholics have grown as more of them have attained the requisite number of A-level passes. By 1982, 13.9

percent of Protestants entered higher education, and 12.9 percent of Catholics (though the latter were more likely to enter the Polytechnic than the universities) (Osborne, 1985: 48). The main difference was that three-quarters of Catholics went to institutions within Northern Ireland, compared with less than two-thirds of Protestants (Osborne, 1985: 48). Catholics predominated in the much smaller group which obtained tertiary education in the Republic of Ireland.

As regards further education, Catholics were as likely to enter this sector as Protestants. Thus, overall, there is both continuity and some change: continuity as regards certain basic features of the educational system, like the gender differential; change as regards the greater opportunities for Catholic children to obtain the requisite qualifications, whether on leaving school or going on to further or higher education. Migration has always played a large role in taking the labour surplus off the market (see Chapter 3), with the emphasis perhaps shifting away from the traditional unskilled or semiskilled building or agricultural labourers (the men) and nursing (the women) towards the more highly qualified graduates.

Why then has the effort of 25 years since the great imbalances were first identified made so little impact on the apparent employability of the Catholic population? There is no clear answer to that. The political climate deteriorated sharply in the late 1960s, and has not improved much since then. The economic difficulties of the Province accelerated rapidly in the 1970s, and this did not help the political climate. Sectarianism, whether in the community at large or at the workplace, was not likely to be mitigated under these circumstances. It is therefore idle to look for some 'educational failure' to explain the excess unemployment of Catholics.

Indeed the danger is that in view of the persistence of high unemployment among Catholics in all areas, and including a high proportion of people with good educational qualifications and professional training, the noticeable efforts of Catholic school managers to change the curriculum towards what is seen to be the right direction could be undermined. The same applies to the relatively ample provision of further education places in all areas, to make good deficiencies still existing among school-leavers, and to supply skills which schools cannot teach. It is much harder to persuade young people to remain in full-time education, and for their parents to fund this continuation, when there is so little evidence that additional schooling and training will avoid the two alternative evils – unemployment at home or the uncertainties of migration, whether to Belfast, the rest of the UK or overseas. We shall discuss mobility in the next chapter.

The verdict on the educational system must remain that which has already been provided by Osborne and his colleagues: that some allegations of inadequacy remain in the air, in the sense that they cannot be totally disproved; but whatever these inadequacies may be, they do not explain the difference in job opportunities. As long, however, as there is a significant difference in educational content and achievement, at any level, between the Catholic and the Protestant communities, there are good grounds for advocating further change. Such demands do not explain, let alone excuse, present inequalities, but they mean that work can be done (or could be done, given funding) to make sure that the preparation of young people for the labour market is the same, regardless of locality and denomination. Only when this has been achieved can any centrally engineered attempt to reduce inequalities by the provision of additional jobs be certain to benefit both communities, and all areas, equally. That this was not so in the past has been, without doubt, due at least in part to differences in educational provision. These differences are much smaller now, but they are not yet reflected, as we have seen, in the structure of the employed labour force.

Finally, all investigators are agreed that of all sectors of social and economic policy-making in Northern Ireland, education is the one which is most closely tied in with the structure of the community in general – more so than in health, housing, social services or technical sectors. The sectarian divide, the centre–periphery differences, the nature of urban and rural communities, the role of Church and family, are all directly reflected in the management of education. The involvement of the priesthood, of employers, of parents, as well as secular political parties, implies direct expression of their traditions and aspirations in the school system. Education is therefore much less amenable than in the rest of the UK to central government direction. On the other hand, significant cuts in public expenditure can affect the educational system seriously and unequally, according to the ability of the local community (where that is involved) to make good shortfalls on current and especially on capital account. Thus the prospects for further changes in the desired direction (for the purposes of this report, greater ranges of qualifications of school-leavers to take account of labour market changes) must be somewhat problematic. Much has been achieved, and much remains to be done.

10

Obstacles to Mobility within the Labour Market

Peculiarities of the Northern Ireland Settlement Pattern

In preceding chapters, the Northern Ireland labour market has been described in terms of acute differences in unemployment, and in industrial structure, between localities. These differences arose within only one of the UK economic planning regions as they existed, at least notionally, until the 1980s. As a planning region, the country is not only one of the smallest in area, but also the smallest in population terms. The surface area is similar to British planning regions such as the Northern, the East Midlands, Yorkshire and Humberside and the West Midlands. In terms of population, it is smaller than any of these; in 1985 it contained about a third fewer inhabitants than East Anglia.

Northern Ireland differs from most of the other UK planning regions in that it has one dominant centre of population and activity, which is also the hub of the system of communications – albeit lying at the extreme eastern margin of the country. It could be said of English planning regions that some are equally monocentric – the South East with London, the West Midlands with Birmingham. The North West has two centres, Manchester and Liverpool, and the other regions are more polycentric. None of these English regions, however, shows such an imbalance of population and economic activity as Northern Ireland.

The Belfast area group which we have used for statistical purposes comprises 45 percent of the population of the Province, and about 60 percent live within a twenty-mile radius of Belfast city. The country's two motorways and remaining railways connect these most populated areas to the capital region, so that about three-quarters of the population lives within an hour's travelling distance from the city. The only other considerable centre of population, Londonderry, is just over 70 miles from Belfast, and that city, with its immediately adjoining settlement areas, accounts for a further 10 percent of the country's population.

Thus the considerable differences which arise can be found in a very small area, compared with British regions whose settlement pattern is much more dispersed. Scotland has a densely populated lowland region, in contrast to the thinly settled Highlands and Islands, and the Borders, but distances are much greater and communications poorer than in Northern Ireland. Wales has a population dispersed around the outer edges of the country. None of the English regions has as strong a contrast between the central and the thinly populated peripheral areas as does Northern Ireland.

What is more significant is that the intraregional differentials in England, Wales and Scotland are less pronounced. Whatever indicator we take, the more heavily populated British regions are relatively homogeneous, and have remained so during the cycles of post-war prosperity and decline. Thus, within the South East region, with a population eleven times that of Northern Ireland, and twice the superficial area, there is indeed a centre–periphery difference in unemployment rates (with Kent, East Sussex and the Isle of Wight having considerably higher unemployment than the inner home counties and the greater London area). But this difference arises over much greater distances away from the centre. Other British regions show even less difference in unemployment rates within the area, whether they are, in total, areas of continued growth or of decline.

In Northern Ireland the differences are larger, and they occur in settlement areas much nearer to each other, and to the main focus of employment. As we have seen, manufacturing industry in Northern Ireland is mostly located in the Greater Belfast area, with minor concentrations round Londonderry and Craigavon, and in those districts of the Northern area group nearest to Belfast. These concentrations accounted for nearly 90 percent of the industrial employment in Northern Ireland, and consequently bore the brunt of the reductions (from 170,000 jobs in 1973 to about 100,000 ten years later).

The relatively small Northern Ireland labour market does not show patterns which we would expect in the British regions. In the first place, given geographical propinquity, one would not expect to see such very large differences in unemployment rates as do in fact occur, as between travel-to-work areas only some 30 miles apart. (We exclude from this generalization the patterns which occur *within* metropolitan areas. For example, in London, such crass differences occur between the multiply disadvantaged boroughs of inner East London, and the outer boroughs only ten miles away, just as within the Belfast travel-to-work area there are similarly strong contrasts between West Belfast and the middle-class suburbs just beyond the city boundaries. There are different causes at work in this type of situation.)

Secondly, if industrial employment is then drastically reduced in those areas which were previously favoured in that respect, we would expect the differences in unemployment rates to *decrease*. Thirdly, if, as we have shown, 'Protestants and others' were employed in many types of industrial production (especially engineering) to a much greater extent than Catholics, decline in these industries should have reduced the unemployment differential.

In fact none of the expected equilibrating mechanisms seem to work. The unemployment differential between travel-to-work areas is persistent (see Additional Tables A4.2 and A4.3). Areas which had important concentrations of manufacturing industry lost proportionately more jobs than areas which had only a small industrial base; but this did not affect their position in the unemployment league tables very much. If it is true that more Protestants lost industrial jobs, relatively, than Catholics, the gap in unemployment according to religious affiliation did not become smaller.

There is thus a measure of inflexibility within the Northern Ireland economy which has meant that the overall picture of disadvantage, always more disturbing than that found in any other part of the UK, remains the same throughout a period of general decline. To this has to be added the fact that, as we showed in Chapter 2, out-migration always removed a considerable part of the potential excess of available labour over jobs. This migration lessened as economic recession hit the whole of the UK, and most other traditional destinations of Northern Ireland out-migrants. Yet net out-movement continued through the period under review, though on a smaller scale.

Reasons for Lack of Mobility

Why then do equilibrating mechanisms fail to lessen the differences within this relatively small labour market? Given that the UK central government also pursued an active regional policy (of trying to steer new employment opportunities into areas of disadvantage) at least until the 1970s, we need to ask the same question for Northern Ireland. When UK unemployment began to rise rapidly in the 1980s, and reached figures only slightly below those for Northern Ireland in the worst-hit areas, especially in Northern England, the question of mobility was raised more frequently, sometimes as a matter of acrimonious political debate.

For the rest of the UK, the answers are fairly straightforward. Apart from the great distances involved (the areas of growth being some 200 to 300 miles distant from the areas of greatest decline), the most frequently cited reason was the scarcity of housing, and in particular price differentials in owner-occupied property. Those who owned their homes in the North could not sell them at a price which would have enabled them to buy a house in the South East or East

Anglia, even if they were sure of obtaining employment.The large distances involved made it impracticable for the principal household earner to commute between the family home and weekday lodgings in places where jobs could be found. This then is the most important reason for the failure of equilibrating mechanisms to function. In addition, it was often alleged that the unemployed in the areas of economic decline did not have the technical qualifications or manual skills required in the regions of growth, but this point has never been proved. Certainly it did not apply to school-leavers who had similar qualifications in all regions; Scotland even provided a better-educated young labour force than England and Wales.

After regional policies had been virtually abandoned, the differential job growth in the most favoured areas accelerated, which resulted in more movement from north to south, and consequently enormous pressures on housing; but this migration was not sufficient to reduce the differential in unemployment rates. Only if economic activity began to grow again (as it did in the West Midlands after 1984) did local unemployment rates fall even without large-scale out-migration (*Regional Trends* no. 22, 1987 regional profiles).

So we must explain what is so different about Northern Ireland that equilibrating mechanisms fail to work, on a much smaller geographical scale. The travel-to-work areas are very small compared with their British counterparts, and the districts of highest unemployment are in most cases only one and a half hours' journey from Belfast. That would mean that daily commuting into Greater Belfast would be neither expensive nor very time-consuming for half the unemployed population, and weekday residence for wage earners, returning home for the weekend, would cost relatively little in time or money – certainly no more than the journeys undertaken daily by a great many commuters in the Greater London area.

When looking for explanations of this phenomenon, we need to look at three main areas – housing, car ownership and public transport – and at other, less visible and probably not measurable, obstacles to mobility.

Housing

When we originally set out to study the structure of employment and unemployment in Northern Ireland, it had been our intention to pay particular attention to the operation of the housing market. If, it would be argued, it was correct that housing was not available in areas where there were relatively more jobs, either because none was being built, or because some sections of the population were barred from access to that housing, then we would have identified a policy area where administrative changes, or the enforcement of legislation, could reduce the differentials in opportunities within the labour market. This part of the investigation did not, in fact, yield useful results.

We studied the reports of the Northern Ireland Housing Executive, their statistics of building, and the various outside consultants' investigations into the workings of the public sector housing market (Birrell et al., 1971; Boal et al., 1976; Harrison, 1981; Singleton, 1981; Northern Ireland Housing Executive, 1978, 1982, 1983; Patton, 1981). We also saw the various reports which looked at the way in which, in the years following the renewed outbreak of open sectarian conflict, households moved, it appeared of their own volition, into areas where they could be nearer those of their own persuasion, where they would feel safer (Boal, 1981). During those years, segregation became more sharply defined. This movement owed little or nothing to economic policies; it did not have much to do with the allocation policies of the NIHE (except that it confirmed the increasing tendency to be careful not to try to mix tenures by denomination when planning new estates). If this phenomenon was amenable to policy direction, it would have more to do with the attempt to promote a political settlement, rather than employment policy. If, as was increasingly found, the working population was afraid both to live in mixed districts, and to cross community demarcation lines when travelling to work, then this might be a serious obstacle to any equalization of labour market opportunities. We believe this to be correct, though we cannot quantify it, and the phenomenon could exacerbate inequalities in employment.

Had we used only the 1981 baseline of Census figures and Department of Economic Development statistics, we might perhaps have had some doubts as to whether housing policies might be in part responsible for high and differential unemployment. However, between 1981 and 1985 unemployment in the Province rose from 104,000 to 122,000. Because travel-to-work area boundaries were changed between 1984 and 1985 (see the Introduction) it is perhaps best to analyse the change only to 1984 (see Additional Table A4.3); this is an increase of 17 percent. The increase in Belfast employment service office area was 21 percent, and in Lisburn 25 percent. In Derry TTWA it was 16 percent. Department of Employment officials have pointed out that the apparently greater change in Belfast, compared with the peripheral areas, is in part dictated by the choice of base years for this calculation: in 1981 the DMC plant at Dunmurry, west of Belfast, had recruited some 3,000 workers, who lost their jobs when motor car production ceased. The smaller the area of analysis, the greater the chance that the opening or closing of a single establishment can affect the issue. The rights and wrongs of locating both largely Catholic housing estates and the De Lorean plant in north Lisburn are not subjects for discussion here, but it does seem that unemployment changes in this case can be directly linked to industrial change. Similar considerations apply to the textile industry areas of the west, and the north-eastern complex of chemical, textile, rubber and engineering industries which have all seen disastrous closures in the last few years. Thus although it is undoubtedly true that if workplaces are barred to one denomination, this exacerbates inequality, the converse is perhaps more pertinent: if industrial jobs are reduced then unemployment will arise, wherever they were located, and housing policy cannot do much to remedy this. The experience of the rest of the UK in that respect has been relevant here. As various studies have shown, the housing prospects of immigrant-descended populations improved between 1971 and 1981 (Lane and Thompson, 1985) but this is not related to the increase in unemployment among the population of the New Commonwealth – even if the housing market does discriminate against them. Unemployment has risen in Glasgow, Liverpool and Newcastle even without discrimination preventing workers moving to where jobs might still be had.

New Building

The total housing effort in Northern Ireland has been relatively much greater than that of the rest of the UK. Until the mid 1970s, housebuilding in the public sector exceeded that of the private sector, despite the substantial subsidies (as well as mortgage relief) being given to private builders, at least until 1976 (the period during which the large Protestant suburbs, especially east and south of Belfast, mushroomed) (Northern Ireland Housing Executive, 1981–83). The subsidies were phased out, and the private sector recovered. The public sector was run down sharply after the mid 1970s, but there has been a change of policy recently, and by 1982 public sector completions again almost equalled private sector sales. The latest available figures suggest that Northern Ireland may be completing about 1,000 houses a year, perhaps 40 percent of them in the public sector, for some time ahead. That is somewhere between 600 and 700 new dwellings per 100,000 population. What this means is easily seen by comparison with the statistics for England and Wales, where housing construction roughly halved in the last decade, with the output of local authorities plus housing associations being reduced to 20 percent of its previous levels and private construction down about one-third from ten years ago. Exact comparisons would in many ways be misleading, but relatively speaking the output in England and Wales was mostly about 300 dwellings per 100,000 population. The comparison is still more startling if we only take public sector output: it is roughly twelve times as high, per caput, in Northern Ireland as it is in England and Wales.

Are these houses built in the right areas, and do they go to the most deserving applicants? That question has been extensively discussed in specialized Northern Ireland literature (Birrell et al., 1971; Boal et al., 1976; Harrison, 1981; Singleton, 1981; Northern Ireland Housing Executive, 1978, 1982, 1983, 440ff.; Patton, 1981), and the verdict must be that although a good deal of bias against the minority was alleged for many years, more recently it appears that allocations have been fairer. There is certainly no evidence from the 1981 Census statistics that Catholics were noticeably underhoused compared with Protestants, except that there were more persons per dwelling; that in southern and western areas, which were more heavily Catholic, the lack of amenities, though fairly small, was more noticeable than in Belfast; and that Catholics were underrepresented in the modern private developments round Belfast. We could find no evidence that Catholics did not have access to modern (NIHE) housing in Greater Belfast. It is also clear from the breakdown of local allocations of the NIHE that segregation by religion was practised to an ever greater degree, and that, as it turns out, a high proportion of Catholics have recently come to live in housing estates where in 1981 there was a great deal of unemployment (Northern Ireland Census 1981, *Economic Activity Report*: Table 12).

It is known that the NIHE has been careful in its respect for the sectarian divide, especially in recent years. Thus the old patterns of settlement by religious affiliation were reinforced: by households voluntarily, though under stress, moving out of neighbourhoods where they formed a minority, whether by exchanging private or public tenancies or even by transferring within the owner-occupied sector; by the allocation of vacancies of existing dwellings in the public sector; and by the fairly rigid divisions observed when allocating newly built dwellings in the NIHE sector. Only in certain middle-class professional areas of Belfast and its outer suburbs has something of a mixture survived, or even increased as the result of sales of properties over time.

It is impossible to say whether the NIHE, or any other government agency, had employment considerations in mind when creating this more segregated pattern. It has always been known that people were reluctant to travel to work if this involved walking or cycling across residential areas occupied by the other denomination; and in recent years this has also applied to public bus services and even journeys by private cars. Employment in the Belfast city centre was therefore very important: the radial nature of the public transport services, and the appearance of a system of private taxi services as part of a 'grey' economy, segregated by denomination, made access to central area workplaces and services relatively easy.

On the other hand, the rapid increase in the decentralized residential and employment pattern of the Greater Belfast area will have reduced the advantage of a transportation system mainly serving radial routes. The more cross-suburban journeys have to be made, the more difficult it must become to find routes perceived as traditionally safe in the context of the old Belfast sectarian settlement pattern. The new peripheral estates built by the NIHE are also segregated, and this causes difficulties when there is no easy access from such housing areas to centres of economic activity. In this context, the location of the De Lorean plant, in relation to the large amounts of new housing mainly for Catholics on the Belfast/Lisburn border, was of great importance. All other industrial estates in that sector of West Belfast are of minor importance compared with the now defunct motor plant.

In the rest of the UK, it is often claimed that the pattern of local authority housing has created a trap for disadvantaged households, especially the unemployed, because exchanges of tenancies within the public sector are generally impossible except within the area of any one housing authority. Thus if the local industrial employment base disappears, as it has done in large parts of East London, there is no way in which public sector tenants can obtain housing in other areas where work might be available (and this problem has been aggravated by the disappearance of a large GLC-owned housing sector). The increasingly complex pattern of journeys to work created by decentralization, especially of the lower-grade non-manual and semiskilled manual employment, with the attendant great increase in the time and cost of journeys to work, cause the concentrations of unemployment to remain coextensive with large public sector housing developments, especially in high-rise blocks.

In Northern Ireland the pattern is slightly different. As in England there is less traditional manual work, and growth is mainly in part-time and low-status jobs for women, which has no effect on male employment chances. The distances from the new housing developments to areas of relatively greater job opportunities, however, are not as great as in London. On the other hand, the perceived difficulties in reaching these potential workplaces (disregarding any question of discrimination) are that much greater. Thus even if, for instance, there were to be a reversal of the decline in jobs in the East Belfast engineering industries, the journey to work from, say, Lisburn, might be considered to be too difficult – notwithstanding the fact that a direct rail connection exists between homes and workplace, a distance of only ten miles (a twenty-minute rail journey).

Thus the one advantage which the British observer finds so encouraging in Northern Ireland – that all public sector housing is under the control of a single authority, and that transfers and exchanges should in theory be fairly easy – is to some extent nullified by the sectarian divide and the reluctance of workers to cross boundaries not shown on conventional maps.

Nor are these matters the only reason why movement within the national public housing sector is relatively small. There are still long waiting lists for new NIHE housing in Belfast, despite the relatively large effort of the last decades. On the other hand, there are many empty houses in Antrim and Craigavon. There is considerable reluctance to move away from one's accustomed neighbourhood, with its family and church ties, and even the promise of a dwelling in a 'safe' area is not usually enough to tempt people away. Evidence on this point cannot be derived from Census sources, but it does rather look as if Northern Ireland workers (and their families) are more ready to migrate across the Irish Sea, or even the Atlantic, than to move within Northern Ireland to an alien environment, even though this may be only a few miles away. It does not seem to have stopped large numbers of people, for instance, from the Falls Road area moving into Poleglass or other Catholic housing estates. Thus there may be a maximum distance (perhaps half an hour's bus journey) which is considered acceptable for a move; anything beyond this may not be feasible. Unfortunately even the Continuous Household Survey, with its questions on intentions to move, does not give us a basis for assessing the importance of this matter.

What is perhaps a much better indicator of the resistance to the type of movement which is common in England among young people, and of increasing importance for adult wage earners with families in areas of high unemployment, is the absence of any sizeable provision of lodgings for single persons in Belfast. Except in the wards around the Queen's University, there is no 'bedsitterland' such as exists on a vast scale in London, and to a lesser extent in provincial cities with a growing service sector. It exists in Edinburgh for Scots, and for the Welsh in Cardiff. We have found no reference in the literature to the reasons for the absence of any recognizable lodger population in Belfast. It may exist in a form in which official statistics cannot establish it: for example, if single workers lodge with close relatives, they would still be counted as one-family households. (This is also true in the rest of the UK, especially in households with origins in the New Commonwealth or Pakistan.) However, given the closeness of local ties, it would indeed be quite surprising to find a sizeable population of workers with homes, say, in the western peripheral areas, spending their working week in Belfast.

In this section, we have, inevitably, had to resort to a good deal of guesswork. Yet the conclusion is quite consistent with the observed fact: that very large differences of unemployment continue to exist as between areas which, by UK standards, are not very far apart.

Housing Conditions and Housing Allocations

Northern Ireland started from a much lower baseline, as regards the quantity and quality of housing, in 1971, so that the relatively much larger effort of the last fifteen years would just about be enough to reach UK average standards. Even at the time of the 1983 housing sample surveys, household size in Northern Ireland was well above UK level, and relative overcrowding more severe. Amenities (baths, indoor WCs) by 1983 had reached British levels of 1977 but the gap was closing. That came about because the rate of new housebuilding and of improvement grants (Northern Ireland *Annual Abstract of Statistics* no. 3: Tables 6.7 and 6.8) was being maintained at a time when both were slowing down in the rest of the UK. All this has been the subject of numerous investigations and recommendations. The latest round of the General Household Survey in Great Britain and the Northern Ireland Continuous Household Survey appears to show that parity of standards has been reached.

In 1969 the Cameron Commission had found overt discrimination against Catholics in housing allocations by the local authorities, and as a result the NIHE had been formed. By 1985 there were very few critics who still maintained that nothing much had changed in the meantime. Roman Catholics had obtained a large share of the public housing stock (proportionately higher than Protestants; see later in this chapter) as well as increasing their share of owner-occupation. Fewer of them were in unfit houses owned by private landlords. This is not a complacent statement. It is merely a reflection of the fact that during the period under review enough new and improved houses came into existence, and were allocated to Catholics, as to make it unlikely that Catholics were unemployed because they were under-housed (if indeed that had ever been the case, though the allegation was made). It could be that too many houses had been built for Catholics in areas where there was no work: but it was housing they wanted, in the places where they wanted it to be. This is in itself probably a reflection of their feeling of insecurity in an alien environment. The same is quite probably true of many Protestants. If that is so, then housing policy as such should not be cited as a cause of differential unemployment. When the building programme was begun, the sharp rise in industrial unemployment in particular could not be foreseen. There is much evidence, in fact, that housing was deliberately sited so as to provide improved access to potential work opportunities.

Could larger households, and higher occupancy ratios, in themselves be the cause of unemployment, lack of access to jobs or lack of promotion? It is difficult to imagine by what mechanism this would come about. It is true that houses occupied by Roman Catholics in 1981 contained an average of just five rooms, and 0.75 persons per habitable room. In the same year, Presbyterians had 5.3 rooms per house, and only 0.55 persons per room. The first difference could be the result of inadequate allocations of larger houses, or poverty; the second is explained by the fact that households themselves were larger (most of the difference being explained by the presence of more children).

Undoubtedly during the 1980s the housing situation may have deteriorated sharply because, despite relatively high building rates, the rate of new household formation has accelerated as the children born during the peak years of the 1960s have married and formed their own households (Northern Ireland Housing Executive, N.D.). There may well be more overcrowding, and it is quite likely that this will affect Catholics more than Protestants – for example, because more married children will have to stay with their parents until they can buy or rent a house. The NIHE lists of completed housing by districts, and their plans for the future, certainly show a lack of adequate provision for larger households (Northern Ireland Housing Executive, 1982). Hardly any dwellings were being built with more than six rooms, and the majority were the smaller types of dwellings with three to five rooms. This conformed with the national policy of concentrating on these houses, of which there was a historical shortage, and in the hope that the elderly (for instance) would vacate larger dwellings for families, and that young married couples would leave the parental home and would in due course be allocated larger dwellings as these became vacant and the young families grew (Northern Ireland Housing Executive, N.D.). This policy may or may not work well, and undoubtedly, given the Northern Irish situation, there are severe constraints on the high mobility rate which would make such a policy work (Boal, 1981). Equally, the programme of sales of NIHE property to tenants, amounting to some 6,000 dwellings a year, reduces the net rate of increase of the public sector housing stock considerably. Unlike the rest of the UK, these sales have not so far actually diminished the stock available for renting (especially when we take into consideration the work of housing associations), but they certainly reduce the rate at which waiting lists can be cleared.

Housing: Continuous Household Survey Evidence

In addition to the Census evidence, we have a certain amount of additional information on the housing situation from the Continuous Household Survey (1983 round), subject to the usual provisos about sampling size, non-response rates etc. (see the Introduction). This showed that nationally, of those who declared their religion, 45 percent of Catholics were owner-occupiers compared with 54 percent of Protestants and 54 percent of the 'non stateds'; 48 percent of Catholics were NIHE tenants compared with 35 percent of Protestants and 36 percent of 'not stateds'; 5 percent of Catholics were private renters compared with 8 percent of Protestants and 36 percent of 'not stateds', plus a small residue of miscellaneous tenures. This confirms earlier estimates, which suggested that the Catholic population was slowly increasing its share of public sector tenures, as well as of owner-occupiers, and that private renting was becoming insignificant (compare Northern Ireland Census 1971, *Religion Tables*, 6, 'Household by religion of head and tenure' with Census 1981, Table 6, 'Household by religion of head and tenure'). In the Belfast subsample of the CHS (643 valid cases out of 2,940 households nationally), 58 percent of Catholics were NIHE tenants compared with 36 percent of Protestants; 36 percent of Catholics were owner-occupiers compared with 45 percent of Protestants. Taking together all urban areas, that is Derry plus certain other towns but excluding Belfast, the proportion of Catholics in NIHE tenures rises to 62 percent compared with 40 percent of Protestants. These figures do not in themselves prove that the Catholics are getting a proportionate share of NIHE housing, or that they were satisfied with what they had, or that these tenures maximized their employment opportunities; at any rate it does not suggest that they were trapped in private tenancies. But even in the non-urban areas (called East and West in the CHS), Catholics in NIHE tenures outrank the Protestants, less so in the East than in the West. Overall, however, it is clear that if anything can increase the access of Catholics to work, it would be transfers or exchanges under NIHE auspices. Given the fact that 87 percent of all respondent households did not think of moving (from 80 percent in the Belfast sample to 91 percent in the West sample), it cannot be assumed that the misallocation of tenures was serious in 1983. Moreover, very few of those who were thinking of moving gave any kind of economic reason for that wish: none, apparently, wanted to leave in order to have a chance of a job as such (as opposed to those who wanted to move because they had obtained a job). On the other hand, of the rather small minority who were thinking of moving, a relatively high percentage were NIHE tenants (17 percent of the 1,154 NIHE respondents); but again it is clear that most of these must have been motivated by dissatisfaction with the size or quality of their accommodation, or its environment, because those were the only sizeable categories among the intending movers.

It is not the purpose of this report to present a critical review of current or past housing policies (including rent policies and allocation practices). Such reviews are numerous, and the NIHE's own annual reports are as apprehensive about their future ability to cope with demand as are the investigations of the critics in many respects.

The Public and the Private Sector

All these anxieties, however, do not shed much light on whether there is any obstacle to mobility of labour within the current practices of house building and allocation, rents or sales. We have already seen that Greater Belfast had relatively less unemployment and better promotion chances, for the Catholic community, than the rest of the country. There is no evidence that NIHE programmes, or the efforts of the private sector, have failed to take into account the fact that this relatively more favourable labour market can only operate if people can move into the Greater Belfast region from the areas of high unemployment. Equally, Belfast still had the worst housing conditions in 1981, and it had by far the largest proportion of closed (bricked-up) properties in areas from which one or other community had withdrawn in search of a safer environment (Northern Ireland Census 1981, *Summary Report*: 4ff., Table 4). In 1981 Belfast itself had 36,600 of the Province's 198,700 public sector dwellings (just under 20 percent, with 21 percent of the population). The Belfast region of the NIHE has a 27 percent share of the NIHE building programme for 1981 to 1986 (the boundary is roughly coextensive with the city), with a further 23 percent going to the (pre-1985) Belfast TTWA, that is, half the new dwellings for the Province (Northern Ireland Housing Executive, 1982). This may not be adequate, but it does not betoken any lack of awareness of the needs. In addition, the bulk of the private sector housing construction in the Province is located in the Belfast TTWA. Of the Province's total housing effort, averaging 10,000 dwellings per annum, half is located in that area and half of that, again, is in the public sector. Thus if access to housing were equal (or proportionate to denominations) there should not be cause for excessive concern – at least against the background of a totally inadequate UK housing effort, stock deterioration, and increasing homelessness.

There is not, of course, 'equality of access' in Northern Ireland any more than anywhere else: Catholics have lower incomes, and therefore their choice of owner-occupied property should be restricted. The evidence, however, points the other way. House prices are very much lower in the Province than in England. Incomes, at least for those who are working, are not that much lower. Thus, in April 1986, average weekly gross earnings for full-time men were over £182, compared with an average of £232 for the whole of England (*Regional Trends* no. 22, 1987). That is, men in the most prosperous part of the UK earned 27 percent more than their equivalents in Northern Ireland. At the same date, a new semidetached house in Northern Ireland cost just over £26,000, and just under £47,000 in South East England (Halifax Building Society, 1986). That means that houses were nearly 80 percent dearer in the affluent South East, and since then the price differential has increased to over 100 percent. So the purchasing power of households with earnings is, in housing terms, much greater in Northern Ireland than in the South East (and, for that matter, than in most other regions of the UK).

Curiously enough, although a higher proportion of Northern Irish households depend on the public sector benefits system, average total household incomes are also not so much below those of the poorer English regions, Wales, and Scotland. (This is in part due to the fact that households are larger, and may have more than one earner, and receipts from child benefit would also be larger.) In the period 1984–85 (years have to be combined to obtain an adequate sample at regional level from the Family Expenditure Survey), UK household incomes were 19 percent higher than Northern Ireland incomes, and South East incomes were 43 percent higher. Thus Northern Ireland households were better placed, in theory, to acquire houses for sale than people living in other parts of the UK.

This is not, however, an entirely theoretical calculation. In the decade 1975–85 UK owner-occupation rose by 19 percent (as a proportion of all dwellings). In Northern Ireland the rise, from a lower base, amounted to 27 percent, and all the current indications are that this process is continuing. And we are here dealing with the period of fastest industrial decline and sharply rising unemployment.

The combination, therefore, of high public sector housebuilding rates, and relatively affordable owner-occupied houses on the market, should mean that the housing system presents fewer obstacles to mobility than we would infer from statistics of unemployment and low incomes.

Conclusion

The Northern Ireland housing effort has for long been the target of severe criticism. There can be no doubt that there is somewhat more unfit housing than in the rest of the UK, that more people are overcrowded, and that a large number of households are dependent on housing benefit to help them pay their rent.

It is not, however, possible to show that housing constraints prevent people from obtaining employment, except in so far as potential workers may feel unable to take up employment because of fears for their safety whilst travelling to work, or because of strong community feelings which rule out distant journeys to work.

Waiting lists in Belfast are still long, but this does not mean that it is impossible for people seeking work to obtain a foothold in Greater Belfast – or at least no more so than is the case in the prosperous parts of England. It is not known with any certainty how far the waiting lists are due to the desire to move out of neighbourhoods which are perceived as an adverse environment to more salubrious areas, and how much to sectarian fears. Nor can we say with any certainty how far the pressure is due to the large number of young people in the household forming stage – and that proportion is still increasing. Again, we do not know how important it is to these young potential households to be near their families. The obstacles may not lie in the volume of existing housing and of housing under construction.

Overall, it is not possible to single out the housing market as a major cause of differential unemployment. People may live, in terms of part of our initial hypothesis, 'in the wrong place', but this may not be because family housing or lodgings for single people are unobtainable. It could be that the desire to move is not as great as high unemployment rates would lead us to expect. This phenomenon is also in the process of being discovered in other economically disadvantaged areas of the UK.

Car Ownership and Use

A further aspect of mobility concerns public and private transport. The inadequacy of public transport is, like housing, the subject of perennial complaints in Northern Ireland. Again this is not the place to summarize, however cursorily, what has been written on the poor state of the railway network, the inadequacies of the bus services, and the community policies which militate against buses but favour an informal taxi service. If people cannot get to potential workplaces, this may be because they do not have cars, because there is no public transport, or because informal taxi networks do not operate. It may also be due to the fact that there is a reluctance to cross certain demarcation lines, whether by public transport, formal or informal, or in private cars.

In 1981, according to the Census (see Table 10.1, Additional Table A8.1, and Figure 24), 40 percent of Northern Ireland households had no car, and that was almost exactly the same proportion as in the rest of the UK. In Belfast that proportion was 58 percent (Figure 25), and

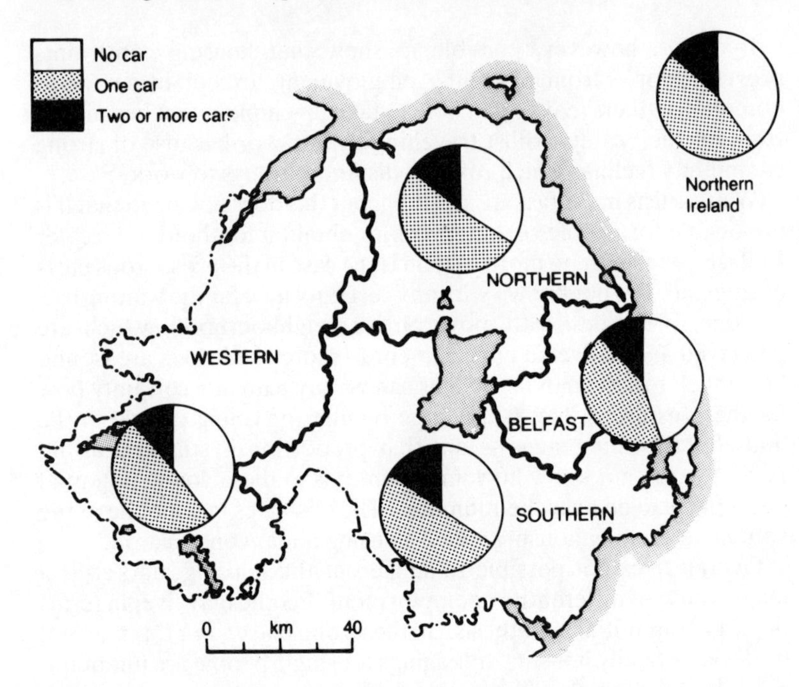

Figure 24 *Number of cars owned per household, by area groups, 1981*

Source: Northern Ireland Census 1981, *Summary Report*, Table 14

Table 10.1 *Car ownership by area group, 1981*

Area group[1]	Total households	Total cars in households	Households as % of all households with no car	with 1 car	with 2+ cars
I Belfast	221,965	154,668	44.6	42.7	12.7
II Northern	72,092	59,161	34.6	50.9	14.5
III Southern	98,542	81,878	34.6	50.2	15.2
IV Western	63,749	47,819	39.2	48.5	12.3
Northern Ireland	456,348	343,526	40.1	46.4	13.5

[1] For definition of area groups see Appendix A.

Source: Northern Ireland Census 1981, *Summary Report*, Table 14

this was the same as in the Central Clydeside conurbation. This implies that car ownership was relatively high outside Belfast, and this is in fact correct: in North Down 30 percent of households had no car, and even in Lisburn only 31 percent. The lack of cars becomes much more serious when we move to the poorer Southern and

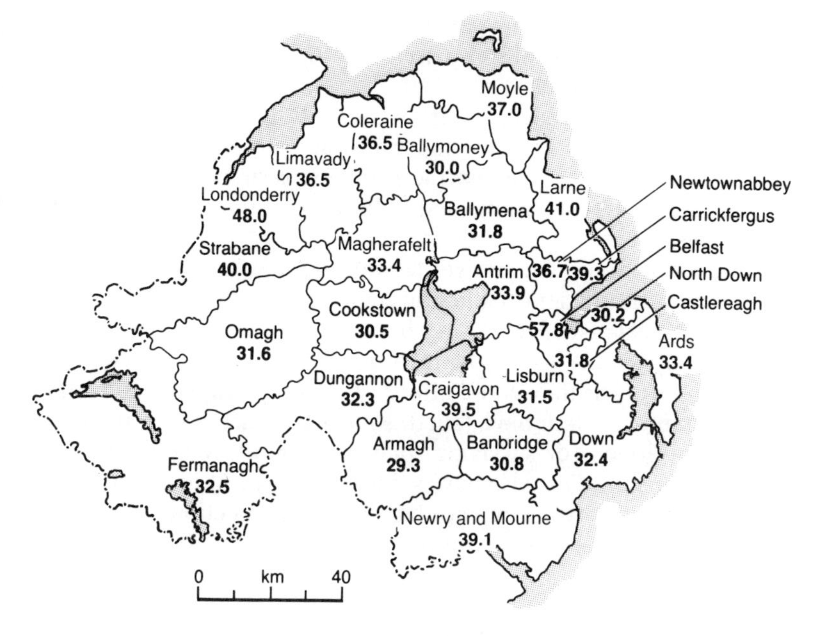

Figure 25 *Percentage of households with no car, by district, 1981*

Source: Northern Ireland Census 1981, *Summary Report*, Table 14

Western areas: in Derry 48 percent of households had no car, in Newry and Mourne 41 percent, and this matters more in a dispersed settlement pattern than in Derry where distances are not so great and some public transport operates. The same applies to Strabane (40 percent without cars), but not to Omagh or Fermanagh (each with 32 percent of households without cars). The figures for rural Northern Ireland are in fact on the whole very similar to British regions with low densities (East Anglia 31 percent, and 40 percent in northern British regions outside conurbations).

By 1983, in Northern Ireland, 58 percent of households had cars (according to the Family Expenditure Survey), compared with 60 percent in the UK as a whole (Northern Ireland *Annual Abstract of Statistics* no. 3, Table 2.6). Slightly later figures using Department of Transport statistics (*Regional Trends* no. 22, 1987: Table 4.2) suggest that, by 1984, the total number of cars per household in Northern Ireland had overtaken the UK proportion, was well ahead of the poorer British regions, and also in excess of cars owned in rural areas like Wales and Scotland. This does not mean that car availability (to household members) was running at a higher level than in the rest of the UK, but at any rate the poverty of the Province was not reflected in car ownership (and this also accords with the possession of some other consumer durables).

Nothing can be concluded from these differences. Agricultural populations everywhere have high car ownership, urban populations much lower levels. Urban Northern Ireland is similar to the poorer urban areas of Great Britain, and Belfast shows the greatest poverty in that respect.

Can people therefore not get to work because they have no cars and there is insufficient public transport? This may be a factor, but there is no real evidence. Poorer households have lower car ownership; unemployed households are the poorest of all, and perhaps they could find work if they could afford cars. All attempts to solve this conundrum in Britain have failed: some of the areas with the most highly developed (or surviving) public transport systems, and the highest fare subsidies, also have some of the highest unemployment rates: neither car ownership nor bus and rail connections create jobs as such, or give access to workplaces to those qualified to occupy them.

Car Availability: the Continuous Household Survey and Census Travel-to-Work Evidence

The only easily available recent evidence on religion and car ownership comes from the CHS of 1983. This survey (based on 2,940 respondent households) puts car availability overall slightly lower than the census, at 58 percent rather than 60 percent, but this difference is within the likely sampling error limits. Within this sample, fewer Catholic households had access to a car than Protestants: 51 percent compared with 61 percent, and 64 percent of the 'not stateds'. In Belfast the proportions were 32 percent, 37 percent and 40 percent, respectively. In all urban areas (Derry and some other towns but excluding Belfast) it was 43 percent, 60 percent and 65 percent. In the non-urban areas the difference was small: 62 percent, 69 percent and 77 percent in the East, and 60 percent, 67 percent and 75 percent in the West. But all told, Catholic mobility should clearly have been lower than Protestant. It would indeed be astonishing if it were otherwise, given the much higher Catholic unemployment rates: this factor alone would account for the difference without considering earnings.

These differences are significant (that is, outside the bounds of any possible sampling error). They will have to be taken into account when summarizing the causes of excess Catholic unemployment: during a period of deteriorating public transport services, the position of households without access to a car, especially in the more thinly populated areas, becomes ever more serious: they may be virtually excluded from the labour market. In parentheses, and in fairness to the efforts of the government, it should be pointed out that the rate of reduction of bus service in the Province is very small compared with the rest of the UK; bus miles run (down to 1983–84) have actually increased over the last decade (Northern Ireland *Annual Abstract of Statistics* no. 3, Table 8.7), though passenger journey mileage has decreased. Staff employed and buses in use have remained stable. Even in Belfast, despite the great difficulties experienced when trying to run public buses in some community areas, the fall in bus traffic has not been serious. Moreover, the rise in the number of unofficial taxis (which necessarily escape the official counts) in all probability means *greater* mobility than had been provided by some of the now discontinued bus services, since they will take passengers to their front doors, as it were, compared with the buses which kept to the main distributor roads on the housing estates. Nor have train passenger miles (important mainly in terms of greater Belfast suburban access to workplaces) been reduced as much as has been the case in the rest of the UK. The unchanged rail mileage – 210 for the whole of the Province, over many years – does in fact connect over 80 percent of the population with Belfast. The required subsidy is very large indeed, and proportionately more than that which is provided for British Rail's 13,000 mile network. Again this is not meant to be complacent: but it does not seem to be useful to look for causes of differential unemployment in the public transport sector as such.

The Overall Modal Split of Journeys to Work

The railway system in any case does not provide an important means of travelling to work. In the Northern Ireland Census 1981, *Workplace and Transport to Work Report*, only Carrickfergus and Craigavon showed more than 10 percent of workers using trains to get to work, and, surprisingly, in North Down (Bangor) just under that percentage travelled by rail. Bus services were significant means of travel to work only in Belfast (where one-quarter of residents used the buses to get to work) and the immediately adjoining residential districts, notably Castlereagh.

Throughout the Province, private cars (or vans) provided by far the most important single method of journeys to work (and the majority of all journeys to work), and this was overwhelmingly true for people living in one district and working in another. In fact in the peripheral areas about 80 percent of work journeys outside the district of residence were undertaken by car.

Of the 514,000 workers analysed in the Census, nearly 400,000 lived in car-owning households, and over 18,000 workers from households not owning cars got to work as passengers or as part of a driving pool. A total of 277,000 people went to work by car, or 53 percent. We conclude, not surprisingly, that the 40 percent of households not owning cars must have included a good many pensioners, as well as unemployed people; but given the importance of bus travel in Belfast, the largest concentration of employment, it is hard to accept the conclusion that lack of a car, or lack of a bus or train service, were major factors in preventing people from working.

The *Workplace and Transport to Work Report* of the 1981 Census unfortunately provides few clues to the extent of interareal mobility. Of the 150,000 persons returned as working in Belfast, only 56 percent were living in the city district. However, of the 85,000 who travelled into the city, 61,000 lived in immediately adjoining districts, and practically none in the peripheral southern and western districts. Outside the immediate Belfast area, only Craigavon supplied more than a thousand workers to the capital. (Some areas of Down are in practice outer Belfast suburbs.) So no evidence exists, as it does in the corresponding tables of the 1981 Census of England and Wales, that appreciable numbers of workers were prepared to travel an hour or more from home to work. There again, it may be a question of custom rather than time and money which limit the distance people are prepared to travel.

Conclusion

We conclude that, although both housing and transport are areas for concern, and although we can construct hypotheses about the connection between lack of mobility and unemployment, the statistics provide little proof. Between 1971 and 1981, and even to 1984, housing improved a great deal, and car ownership grew. This did not moderate the rise of unemployment. In Northern Ireland by 1985 there were more than 40 percent more private cars than there had been in 1971, and unemployment trebled. Such lack of correlation can be found in Britain, even in areas of high unemployment. What is involved are changing real costs, and changing perceptions and priorities for individuals.

The same is true of housing. The total housing stock is hard to estimate accurately for any one year, but it is likely that it rose by 10 percent net between 1971 and 1983 in respect of *occupied* dwellings, while at least a further 30,000 dwellings stood empty. Average household size fell, and room occupancy declined. Increases in stock were concentrated on the areas of the relatively best economic opportunities. But this means neither that housing always went to those in greatest need, nor that the system operated so as to maximize the employment opportunities for those who were out of work.

So we are left with the imponderables: the obstacles to mobility which are known to operate but which no investigation we have seen has sought to quantify. The fear of living in areas, or travelling to work through them, which are felt to be hostile to the Catholics in particular, is real enough: so, even if jobs were allocated proportionately to the denominations, many might not feel able to take them up. However, we do not wish to overstress this factor, apart from the difficulty of identifying it accurately: as our analyses have shown, Catholics are badly underrepresented in employment even in their own areas, and they tend to hold the less remunerative jobs, with fewer prospects, even where they do have work. This can be connected only tenuously with their geographical mobility.

As we showed in Chapter 2, mobility within Northern Ireland was quite high in the years before the Census, as measured by the movements between districts. However, much of this was movement across administrative boundaries, but over very short distances only: most of the intercensal loss of Belfast district is represented by movements to five adjoining local government districts. (In one year alone, 5,000 people were recorded as having moved from Belfast to the districts of Ards, Castlereagh, Lisburn, Newtownabbey and North Down.) By contrast, only about 150 people had moved into Belfast from the four most disadvantaged western peripheral districts. Again, this does not show whether or not mobility was great enough to maximize employment opportunities. And, as our references to the Continuous Household Survey showed, given the high rates of unemployment, the proportion of those thinking of moving was relatively low. Here we are clearly in the realm of highly intangible factors operating within the Northern Irish communities. We cannot comment further on these.

11
Overview and Conclusion

The detailed statistics in the text and the Additional Tables, and the interpretation placed on these, are the results of nearly four years of intensive data collection and subsidiary calculations performed on these data. The objective of this exercise was to try to see whether we could, by the analysis of Census data, survey results and the records of specific arms of government, confirm or refute a series of hypotheses from which this inquiry began (see the Introduction).

As was only to be expected, the outcome is not wholly conclusive. A number of important, but not easily quantifiable, factors must still be investigated by other methods.

The Agreed Facts

From the start, there has never been any dispute about the outcome of the workings of the labour market during the last 25 years in Northern Ireland. Unemployment in the Province has always been high relative to the rest of the UK, and it has remained higher than that recorded in any economic planning region in Great Britain. Relatively speaking, the unemployment rate is no longer so much larger: overall, for instance, it is no longer twice as high as the UK average, as it had been until 1975. A number of subregions in Britain now have unemployment rates approaching those of Northern Ireland (*Regional Trends* no. 22, 1987: Chart 8.18 and Table 8.21).

Belfast (the core) was always relatively better placed than the south and the west of the Province (the periphery). Roman Catholic unemployment has always been higher than Protestant unemployment, and it has never been disputed that this is in part a function of the fact that there are relatively more Catholics in the economically weaker parts of the Province. Within the labour force it has always been known that the Catholics were underrepresented in the managerial and professional grades.

In the last ten years, unemployment has grown rapidly in all parts of the UK. However, towards the end of the period during which this report was being written, there were signs of the rise in unemployment levelling out and possibly, after 1986, beginning to fall. Northern Ireland appears to have shared in this movement, at least to the extent of recorded unemployment not rising much further. It is difficult to be sure about this, because methods of reporting on the state of employment, and eligibility criteria for benefit, have changed so much that the trends may be apparent rather than real. The relative position remains the same: Northern Ireland has the highest unemployment rate of any UK planning region. Moreover, whereas in the UK as a whole a complementary figure, that of all employees in employment plus the self-employed, rose fairly steadily, the Northern Ireland figures both for employees in employment and for the self-employed have remained static.

There has been no dispute about the main cause of increasing unemployment: the differentially greater rate of closure of manufacturing industry plant in the Province. Firms which had often been established on the basis of heavy subsidies and large outright grants, in pursuance of the most massive regional equalization policies adopted by successive British governments anywhere, ceased to trade. Public expenditure, though subject to some cuts, has been on a relatively much larger scale in Northern Ireland than in the rest of the UK, and this sector became increasingly important, not only as an employer overall, but as an area with the relatively greatest stability over time; it made the largest contribution to employment in the declining regions of Ulster, and it was the principal hope for Catholics as a source of employment.

We are not here making any pronouncements on these policies. Others have commented on expenditure on security as a source of employment; on the wisdom of financing branch plants of multinational firms in a period of recession to enable them to operate in an area which had little natural attraction for them; and on the extent to which public sector employment growth really did provide more and better opportunities for Catholics (compare Bradley et al., 1984).

Some Subjects for Conjecture

It might be desirable to break down the rise in unemployment into its component parts: in particular, the gross number of job losses, plus the increase in labour supply, set against the new jobs created. To some extent we tried to do this with our overall labour market balance sheet in Chapter 4, but in fact any such computation must remain somewhat speculative. It is obviously true, given past fertility differentials, that the increase in labour supply played a relatively larger part in the total labour market failure than it did in the rest of the UK, even assuming that the rate of job loss was the same in all parts of the country. Such a division of the total outcome into only two contributory types of factor would, however, be a gross oversimplification, especially if it were used to blame past fertility rather than industrial decline for the present state of affairs. The most that could be said would be that the effect of the collapse of manufacturing industry and construction would have been less disastrous if fewer school-leavers had reached the labour market during this period.

It might then, however, be said with equal justice that the effect would have been less if out-migration had continued at the rate of the boom years. Then, high demand for labour in the countries to which Northern Ireland workers traditionally migrated, plus for a short time the Republic of Ireland, combined with an increase in job opportunities in several sectors of the Northern Ireland economy, produced relatively low unemployment rates, *despite* the fact that the labour market entrant cohorts were increasing year on year.

Another complicating factor, when considering the effects of high labour market entry, is the fact that to some extent employment opportunities, in the context of a functioning welfare state, are directly related to population growth. Health care, education, housing provision and social services, were supposed to grow *pari passu* with population increases, and the rise in child population would be particularly important in that respect. To some extent, also, the growth of the private service sector can be linked to population growth.

Therefore a simple linkage between labour market entry and falling industrial employment to 'explain' the rise in unemployment, is inappropriate. We offer a mere statistical balance sheet to show how many extra jobs would have had to be created to ensure high levels of employment.

The Persistent Differential

The main subject of this investigation, however, has been the attempt to explain the differentials which are observable, rather than total changes. As we stated in the Introduction, we had to examine two opposing sets of contentions, given the consensus about the fact of differential economic advantage. On one hand there was the claim that it was Catholic population growth which caused Catholic unemployment, together with the assertion that Catholics tended to have too few skills and qualifications for the jobs that were open; that they lived in the wrong places, that they were unwilling to move to where work was, and so on. On the other hand there was for a long time the Catholic community's own belief that their lack of opportunities was due to discrimination against them at the workplace, and their fear that they could not move, even if they wished, to areas of better opportunity because the safety of their families would be jeopardized.

A few years ago, one would have added discrimination in housing allocation to the factors causing differential unemployment. This charge, however, looks increasingly unlikely to be a serious factor since the establishment of the NIHE, and the patterns of tenure we examined in the preceding chapter. Residual anxieties on this score may, however, remain; nor has the fear of discrimination at the workplace noticeably diminished.

The Role of Migration

All parties are agreed that the problem was aggravated by the reduced rate of out-migration from Northern Ireland to Great Britian and overseas, though the extent of that migration has recently been in dispute (see Chapter 2). It is questionable how far the problem has been aggravated by an accelerated outflow into the resurgent labour markets of the Irish Republic in the early 1970s, and the subsequent reflux when unemployment mounted in the South. It is not even known how much of the excess unemployment in Ulster is directly due to rising unemployment in Great Britain: that is to say, we do not know how many people from Northern Ireland were added to the local unemployment total as a direct result of the recession in Britain, not only preventing traditional labour migration, but actually exporting unemployment.

In the summer of 1984 there were 22,000 unemployed male construction workers claiming benefit, but in ten years employment in the Northern Ireland construction industry fell by only 16,000. Allowing for natural wastage including early retirement, the presence of 22,000 persons who were registered as construction workers suggests that many of them were returners from Britain. It is less easy to make similar calculations for the rest of the economy.

Perhaps we should also stress that unemployment figures are notoriously unreliable in the construction industries because of the large number of self-employed ('lump') workers who may or may not be included in the self-employed counts, and large amounts of casual labour which may not show up in any statistical table. If we consider migrant building trade workers, they may easily escape all Census counts, insurance statistics, the Labour Force Survey and so on. However, the figures as published strongly suggest that the excess of registered employees over employment loss in the construction industry gives us a valid clue as to what was happening: if insurance figures are somewhat chaotic in this industry, it is unlikely that the number of registered unemployed *overstates* the discrepancy.

The Role of Differential Demographic Structure

So, what do our analyses tell us about the position of the Catholics? There is no doubt that in the past, and even recently, Catholic fertility has been higher than Protestant fertility. Both denominations have recently shown signs of a sharp fall, and the gap appears now to be closing, as it has done elsewhere in the world. However, the young people who entered the labour force in the 1970s, and are still doing so now, are the products of a period of very high fertility, culminating as elsewhere in the 'bulge' years of the mid 1960s. This alone would make it more difficult in a stationary (let alone a declining) labour market to find jobs for all potential entrants, but other factors also play a role. The older male age groups, those whom the entrants would normally replace, are relatively thin on the ground. That is so because Catholic emigration, in the years of heavy demand for low-skill labour in Britain, was greater than Protestant emigration. Even of those who remained behind, however, a high percentage were not working at the time of their supposed retirement. They had been unemployed, or long-term sick, or had taken early retirement. So, as we saw in Chapter 4, the imbalance between potential labour force entrants and leavers was exacerbated by the employment structure and status of the older generation, as well as the numbers of the young.

As we demonstrated, this demographic imbalance is a general problem, but it is found in its most extreme form in Northern Ireland, among Catholics, and in certain districts. There are many European countries where similar imbalances occur, first because the birth bulge of the 1960s was nearly universal, and secondly because for various reasons the retirement age group was, and is, rather small: for example, in Germany the war took a heavy toll of the men who were due to retire in the 1980s. In areas where for any reason the average age of the existing labour force is young (for example, in many of the growth areas of Great Britain), there is in any case an age imbalance problem in the labour market: but if there is growth, then the young entrants can still be absorbed. If there is stagnation, or decline, then youth unemployment is more or less inevitable.

This is not an explanation of high Catholic unemployment rates, but it puts in context the role of fertility in producing excess unemployment. It may be calculated, very roughly, that if the cumulative reproductive performance of the Catholic population in the fifteen years before 1981 had been the same as that of the Protestants, there would be about 60,000 fewer Catholic children under 15 or, say, 4,000 fewer entrants into the labour force each year (males and females). If this were the case, it would actually mean *lower* age-specific fertility rates for Catholics than for Protestants, because their general age structure is younger. Would such a reduction in the Catholic child population mean that unemployment would be less unequal?

The answer to that question cannot be given unequivocally. It is theoretically possible that if, for several decades, Catholic fertility rates had been the same as Protestants' rates; if Catholic emigration had been proportionately the same as that of Protestants; if access to the labour market were equalized geographically, and no discrimination existed at the workplace; then Catholic unemployment rates would now be the same as Protestant. Such speculations have purely theoretical interest. Even if there had been greater similarity between the two main religious groups in respect of these variables, the two populations would not have been equal in all other ways. Their social and economic structure has been evolving along quite different lines over a long period.

The best answer to the problematical question, however, is in the quite recent past. In 1971 unemployment among stated Catholics (with far fewer 'not stateds'), as a proportion of all those economically active, was 14 percent; for Protestants and others it was 6 percent, for Catholic men it was 17 percent. So the chances of the Catholic working population have historically been lower, in the labour market, even before the recent demographic changes. In 1971 Catholic children under 15 formed only 38 percent of their age group, compared with the 46.5 percent we found in 1981, so that the excess unemployment in the 1980s cannot be causally linked to the larger Catholic child population in 1981. Those who attempt to make the connection seem to overlook the time lag between birth and labour market entry – a period which becomes longer as children stay longer in full-time education. The further we go back in the post-war evolution of the Northern Irish population, the smaller the share of Catholics in both the child and the adult working age groups; yet the excess unemployment was always there. This is another reason why it is not legitimate to blame Catholic fertility for the present heavy imbalance.

Spatial Distribution

What about the spatial distribution? In general, unemployment is higher in peripheral areas than in the core. This applies whether that periphery is predominantly Protestant, as it is over a large part of the

Northern group of districts, or Catholic, as it is in the South and West. It applies within Belfast, however, that is, at the centre of all economic growth, as between the largely Catholic West Belfast sector and the rest of the city.

The same explanations clearly cannot apply to both cases. Peripheral area unemployment has always been higher than that in Belfast, and it has become worse because of the deteriorating industrial situation. Belfast has also suffered severe industrial setbacks, but still the same sort of unemployment differentials remain. In the past, Catholic populations have moved out of the declining rural areas, and out of Derry: to the East, to the UK, to the Irish Republic. Those that remained had, it would seem, a reasonable prospect of being at any rate no worse off than they had been, given governmental industrial support policies and the growth of the public sector. In fact stated governmental policies of industrial support, of infrastructure investment, and of current expenditure allocations (including housing) were specifically designed to ensure that the population stayed where it was.

Regional Policies and Mobility

There is no hard evidence to show whether these policies did in fact have any effect on mobility. At any rate it is clear that the claim in hindsight that Catholics have higher unemployment rates because they live in the wrong areas is particularly wide of the mark: even if there had not been some reluctance on the part of Catholic workers to move towards the core of economic activity, governmental policies were designed to stop them from doing this.

The situation in Great Britain is different. There has never been any particular reluctance on the part of workers to move across the country in search of work, whether it was the migration into the new industrial north in the eighteenth and nineteenth centuries, or the southward and eastward drift in the twentieth century. Government policy tried to prevent this movement, both because of the adverse effects on the regions of out-migration, and because of the housing shortages and other aspects of congestion, which were characteristic of the booming Midlands and South East. Only in the 1980s, when regional policies were virtually abandoned, did the natural tendency of workers to move in search of work, and the government's encouragement for them to do so, coincide. By this time, however, the mounting difficulties in obtaining housing in the more prosperous regions had put much more severe obstacles in the way of accelerated migration.

The Northern Ireland case, therefore, remains sharply differentiated from the rest of the UK. As our tables show, net emigration from the disadvantaged areas has in fact continued at quite a high rate, but it does not seem to have made any difference to their excess unemployment. (One could say, in theory, that had they not moved out, in their tens of thousands, unemployment would have been higher still.) As for West Belfast, it is absurd to say that its inhabitants should have moved out in order to avoid unemployment. (We have not analysed Belfast unemployment by ward and religion separately, as this seemed both unnecessary and too dangerous in view of the high differential non-response rates: but just taking the eleven distinctly Catholic wards in the city, none had less than 35 percent male unemployment, and several over 50 percent.) So their immobility, had this been a fact, would not have been a factor in preventing them from getting jobs. In the event, a great many Belfast Catholics moved out of the city, as did Protestants: the former mainly to NIHE developments, the latter to private estates. This did not redress the balance of unemployment.

Skills and Employment Chances

The question of skills is much more difficult to analyse in any conclusive manner. Professional qualifications are the easiest to deal with, because they are separately recorded in the Census under the appropriate headings. In 1981 fewer Catholic men were professionally qualified, relatively, than Protestant men, but more Catholic women. Taking the most common headings, teachers and nurses, we find that 38 percent of Catholic qualified men were found in these two occupations, and 75 percent of women. For Protestants the figures were 21 percent for men, and 70 percent for women. In the small categories of social and business studies, and in other arts and humanities subjects, Catholics and Protestants had about equal proportions within the qualified workforce. In the vocational fields, however, and in the pure and applied sciences, there are large differences, with 42 percent of qualified Protestants in fields like architecture and professional engineering, and only 25 percent of the Catholics (see Additional Table A7.1).

In 1981, total employment in all these fields amounted to only 15 percent of all workers. Thus the distribution of qualifications can be equated with the absence of Catholics in the professional and scientific grades (excluding teaching and nursing), but it throws no light on overall unemployment. The imbalance in that part of the labour market must help to explain the scarcity of Catholics in leading positions in the economy. It is not, however, a surrogate for information which would explain the alleged underqualifications of Catholics in the middle-rank white-collar and manual occupations.

A number of attempts have been made to measure the relative lack of subprofessional qualifications among school-leavers in the two main denominations. R. D. Osborne (1985) has shown that, at any rate in the period 1979–82, there was a continued shift towards 'useful' subjects among Catholic school-leavers. There was still, at the latest date, a tendency for grammar schools in particular to concentrate somewhat more on humanities subjects, and this was specially marked in the Catholic grammar schools. Unfortunately, such detailed analyses are not available for the 1950s and 1960s, when most of the present labour force left school. (For the most accurate statistics available, see Osborne and Murray, 1978.)

Certain subjects (for example, technical drawing) were virtually absent from the school curriculum for both denominations until recently. In other words, the whole Northern Ireland labour force was underqualified in the perspective of the labour market twenty years after the date of the O-level analyses which underlie the general judgement on the suitability of the school-leavers for future employment.

A recent analysis of labour force qualifications comparing all UK regions with each other, and based on the Labour Force Survey, does indeed show that in 1985 the Northern Ireland percentage of persons in the workforce without any qualifications at all was the highest in the UK (41.7 percent), but that was only just ahead of the West Midlands, where there were 40.7 percent of such workers. The UK average was 34.7 percent. Northern Ireland, however, scored above-UK percentages for some categories, for example 'other' (non-degree) higher education, and completed apprenticeships. The Labour Force Survey does not distinguish respondents by religion, but the figures are consistent with the general picture presented in Chapter 8 (*Regional Trends* no. 22, 1987: Table 8.9).

Similar statements are made in relation to the labour force in the disadvantaged regions in the rest of the UK, and they are not 'explanations' of unemployment. For this reason it is misguided to attempt to use, for instance, regression techniques to show the connection between unemployment, lack of skills and religious affiliation; multiple regression does not explain the direction of the causal nexus (Doherty, 1980). The *association* between these critical variables is not doubted; it does not, however, throw light on the operations of the labour market, especially in the past.

Differences, then, have existed in the past, and continue to the present in an attenuated form. They are deeply rooted in the structure of the Catholic and Protestant communities; in their different attitudes towards the state-provided education system; in the nature of the Catholic voluntary schools as well as the predominantly Protestant grammar schools which, in the past, constituted the educational elite sections in Northern Ireland.

No value judgements are involved in making these distinctions. For the three decades following the 1944 Education Act, in Great Britain, there was a great deal of emphasis on the need to provide a broad-based education for all children. It was axiomatic that vocational instruction was to be only a small part of the curriculum in the majority of schools in the tripartite system, and that high standards of literacy in particular should be inculcated into the whole population. If they needed technical skills these would come later, from the further educational system, whether on a full-time or part-time basis, and from apprenticeships.

Since these beginnings the change to comprehensive schools, at least in England and Wales, has further emphasized the need for a common curriculum, in which the less academically gifted children would still obtain an insight into the wider cultural heritage, and those destined for tertiary education would acquire practical skills.

Most recently, the emphasis has swung back again; and it has become fashionable to demand that every pupil should receive vocational instruction at school to fit her or him for immediate useful employment. The arts and humanities were relegated to a relatively less important status in this latest turn-about.

These are tentative conclusions on the subject of qualifications: much of this cannot be quantitatively tested. It is possible, then, that good overall educational attainment on the part of Catholic school-leavers did not fit well into the economy of the period after 1975, and that certain technical skills entered the Catholic curriculum fairly slowly when compared with the fast rate of change within the labour market.

However, as we have seen, high Catholic unemployment featured long before the old economy collapsed and the new skills were demanded more widely. This fact, when put side by side with the fact that qualified Catholics also had relatively higher unemployment rates than Protestants, rules out the possibility of blaming unemployment largely on the Catholic schools system, as was demonstrated in Chapter 9.

As time passes, and fewer children enter the labour market each year, of whom a higher proportion are qualified, and as Catholic educational patterns become more like those of the Protestant voluntary schools (as well as the state system), the old explanation will count for less and less. The most recent Continuous Household Survey analyses do not suggest, however, that the

Catholic/Protestant differential in unemployment is lessening significantly. Other forces must be at work.

When we turn to the structure of secondary, further and higher education in 1981 (Tables 9.5, 9.6 and 9.7) we cannot see any such difference. Catholic labour market entrants in the 1980s are as likely to be qualified for the more professional and highly skilled jobs as Protestants; whether they obtain employment cannot be predicted from our figures. Overall, as we tried to show, a higher proportion of Northern Ireland youngsters obtain post-secondary education than is the case in England, and the Catholics have a slightly higher share of this advantage than Protestants. Again, this does not exclude the possibility that relatively too many of these educated young people are likely to become teachers or nurses or social workers and do not obtain the higher-status posts, especially in the private sector. Whether that is from choice, whether it is due to discrimination, or whether it does have something to do with home areas, we cannot yet state with any certainty. In the structure of further education in the Province (most relevant for more highly skilled employees, for the recruitment of junior management and supervisory grades) there is nothing in the area distribution to suggest that the people with relevant qualifications are not being trained everywhere.

The Roots of Prejudice

In Chapter 10 we looked briefly at housing and transportation, and concluded that there was very little in the available Census and survey data to suggest that the provision of housing, especially in the public sector, and the pattern of car ownership were much of an explanation for differential employment and promotion chances, though certainly it does not help if unemployed people have no cars and live in remote areas.

The main conclusion from this survey of the traditionally cited causes of Catholic unemployment, and poor showing in the employment status classifications, is that the popular current explanations (which blame the Catholics for their own misfortunes and were quoted in the Introduction) have little or no validity. Nevertheless, it could be true that if Catholics 25 years ago had produced fewer children, had moved away from their home areas to an even greater extent, or chosen a different educational route, the picture might now be somewhat different. As an explanation of the present situation, however, statements of what might have occurred, had people or institutions or governments behaved differently a generation ago, are unacceptable speculations to social analysts. When we say certain trends *might* have occurred to a greater or lesser extent, if patterns of behaviour had been different in the past, we are merely guessing.

It should be noted that it is a common phenomenon to attribute hostility towards a minority (anti-semitism, racism generally) to the *size* of that minority at any given time, or in an area, or in relation to a particular commercial activity, or in a professional group. The *fear* of any existing proportion growing further (whether owing to alleged excess fertility, or to favouritism being shown to members of the group), is then propounded as a rational excuse for persecution, or discrimination. The implication of this attitude is that the minority in question could reduce the friction existing between it and the host community, or the majority in any given population, by limiting its numbers. This is obviously not the case, because such hostile attitudes exist whether the minority in question comprises 1, 5 or 25 percent of the population.

If, then, most of the most obvious technical reasons why Catholics suffer from worse unemployment in Northern Ireland, and generally occupy low-status jobs compared with Protestants, are not applicable, we must turn back to our core chapters on industrial and occupational structure in the Northern Ireland economy to see if they yield more precise answers to the questions about Catholic disadvantage.

The Employment Structure in 1981

In 1981 Catholics were heavily overrepresented in low-skill, low-status, low-pay occupations. A high proportion of the women were in part-time service jobs. Whole important areas of industry (especially metal and electrical manufacturing, engineering, shipbuilding) employed few Catholic workers. Apart from the security jobs (the category includes police, firemen and ambulance crews, and private security firms), where Catholics were noticeably underrepresented, they scored better in the public service, mainly in teaching, nursing and social work, and in the low-level ancillary jobs connected with the public sector.

The position in the public services has been investigated generally by the Fair Employment Agency (FEA, 1983), and in greater detail in various sectoral reports published by them between 1982 and 1986. These highlighted both the less than proportional representation of Catholics in the sectors investigated, and progressively decreasing representation as we go up into the supervisory and managerial grades. We will not comment further on these investigations, which led at least to an attempt to rectify recruitment and promotion practices.

As regards the security services in the narrower sense (Royal Ulster Constabulary, private security firms) it hardly needs pointing out that, given the prevalent identification of the nationalist elements as the main sources of violent crime, it is likely that rank-and-file recruits (let alone the officer grades) should be predominantly found from among the Protestant population, and that Roman Catholics might find it difficult to make the decision to join. Though attempts were made to find minority entrants to the police force in particular, this has not been very successful, and the position is reflected in the figures we gave in the relevant chapters.

Whether we look at the situation nationally, in area groups, or at district level, the picture is monotonously similar. Even in the more heavily Catholic areas, Catholics got less than their proportional share of managerial, professional and supervisory positions, men and women alike. In predominantly Protestant areas, their share was even smaller. Everywhere more of the Catholic men are economically inactive (early retirement, permanently sick) and, of those economically active, a far higher proportion are unemployed, in some cases in the proportion of three to one (relative to their share of local population). Here and there we have exceptions, which have been noted: they do well as managers of retail establishments, for instance.

The result of this exhaustive investigation of the structure of the labour market is quite clear. Catholics stand less chance of being employed, wherever they live, except in those occupations where denominational traditions ensure that they do get work: in their own schools, and as nurses. Whether, outside these two professional groups, the relative underrepresentation of Catholics is due to the lack of relevant skills or technical qualifications, cannot be stated with certainty. As we have shown, the educational developments (at any rate of recent decades) make it less and less likely that lack of appropriate qualifications can be cited as a reason for not recruiting Catholics seeking work.

History as an Autonomous Factor

Disregarding then the special case for the public service categories, important as they are, what can we say about the structure of employment generally? One simple way out is to accept that history plays a much larger part in determining present-day social and economic structure in Northern Ireland than elsewhere in the UK. On that score, we could just say that Protestants have always had the better jobs, have been more highly paid, have lived in better houses, and have reached the higher social status groups (Hepburn, 1982: 23ff.). This is tantamount to saying that over a period of 80 years, through several changes in the administrative framework, and through a long period of growth in education, the relative positions of Catholics and Protestants have not changed. To accept the strength of historical forces, however, is merely to try to escape from the more fundamental question: if that long-established trend is so strong, could not methods have been devised to lessen the disadvantage of the minority population? Has all the legislation enacted, and the attempt to enforce it, been nugatory?

We have tried to show that the causes for the continued inequalities in the employment market do not, by and large, originate in the structure of the Catholic community itself. It would be convenient for those who are being blamed for the present inequality between the denominations, if the long history of the matter had permanently conditioned the Catholic population into the role of an underclass – a term that is now coming into fashion again in the United States. If the minority in Ulster acts out this role, the position could be accepted as unalterable. In the general social analysis, this is known as 'blaming the victim'. In Northern Ireland, however, the law (and, be it said, the administration) does *not* accept that this historical conditioning into the acceptance of a subordinate role needs to be a permanent feature of Northern Irish life. Moreover, some progress has been made, as successive reports of the Fair Employment Agency show. The ordinances relating to 'contract compliance' are being enforced in the case of public service contractors, to an increasing extent. More Catholics are now said to obtain promotion into managerial positions. There was, indeed, not much progress between the 1971 and the 1981 Censuses, but these were years of quickening economic disaster – not perhaps a propitious time to break with the tradition of centuries.

The Adverse Effects of Recession

The expansion of the public service sector ceased in the late 1970s, and there has been contraction since that time. Industry shed a sizeable part of its labour force, and the plants that remained were subject to the same process of rationalization and amalgamation as that which occurred elsewhere in the UK. Managerial positions were therefore proportionately reduced. Only in the service sectors was there any considerable expansion, and, within that, the great majority of vacancies were for unskilled, low-paid and part-time workers.

Having allowed for this change in the context of the labour market situation, we are still a long way from showing that the position of the Catholic population is due to external circumstances, as it were, or to the after-effects of historical developments.

There remain only two possible explanations, neither of them easy to quantify. One of these relates to the attitudes of the Roman Catholic working population, the other to that of the Protestant employers.

The Residual Explanations

We have shown in various parts of this investigation that despite apparently quite a normal propensity on the part of the population to move house, mainly across the boundaries of adjoining administrative districts, and despite the historically high rates of migration across to Great Britain (and back again), to overseas countries, and to the Republic of Ireland (and back again), we have not been able to demonstrate that this betokens a willingness to take the risk of going to live in a hostile environment. On the contrary, the outcome of the considerable movements within Belfast, and into the adjoining districts, over the whole period since 1961, has been a more highly segregated pattern of residential settlement. In as far as these movements were due to the activities of the NIHE, this segregation was desired as well as being facilitiated officially. As regards movement into owner-occupied property, such voluntary segregation is more difficult to prove. The religious composition of the districts adjoining Belfast, to which the majority of new owner-occupiers moved, shows a large preponderance of Protestants. Lisburn, the only district with a sizeable Catholic population, owes this entirely to the new housing estates adjoining the Belfast boundary.

Thus there has been no tendency for Catholics to move into areas which might have been advantageous from the point of view of finding work, if it meant forsaking the safety of their community boundaries. Only if they had already obtained professional, managerial and higher administrative positions would they move into the middle-class suburbs.

What of the propensity of the Catholic workers to seek employment in Protestant-dominated plants and service establishments? Here we enter the most difficult part of the analysis. Briefly, Catholics allege that they do not apply for vacancies because they will be discriminated against – either by not obtaining work at all, or by being denied promotion within the firm (or public service sector). They are also said to fear journeys to work that take them across hostile territory.

The Protestant response to this is either the classical one stated at the beginning of this book, that Catholics are unsuitable for many jobs because of their lack of qualifications or because they live too far away from the centres of expansion; or the alternative one (and this surfaces frequently in the FEA investigations) that there were no Catholic applicants for the vacancies which arose. Here the argument becomes circular. To this one might add the finding of all investigators that, in the private sector at least, vacancies are filled, not by being advertised or through job centres, but by passing the information by word of mouth within the families already working in the plant. This practice is known, of course, in all traditional British sectors like shipyards, coal mines, iron and steel industries and others with a strong localization factor. That is, where plant and settlement are part of a whole community, there is not much room for outsiders. Sons following fathers (and in the case of textiles, daughters following mothers) are part of a well-established tradition in industry, and not only in the British Isles. In as far as much of Northern Ireland industry was of this traditional type (textiles in the west of the Province, engineering in East Belfast) this is not really a surprising finding.

Why is Northern Ireland Different?

Yet there is a fundamental difference. Whereas this strong localization and local recruitment is to be found elsewhere, it does not usually take the form of keeping out newcomers merely on the grounds of religious affiliation. Indeed, where there is growth, newcomers are welcomed by employers and, even if they experience some initial hostility by the existing workforce, integration eventually takes place. Thus, between the two world wars, tens of thousands of Welsh miners came to the West Midlands to work in engineering trades. After the Second World War, Caribbean immigrants were recruited for the basic-grade jobs both in industry and in public services. In France, Polish workers were attracted into coal mining. In Germany, Turks came in their millions to fill vacancies in production and services. These minorities were not usually treated well, at least initially; only in the USA were the immigrants rapidly absorbed. So far from being denied a job on the grounds of race or religion, workers were actively recruited. This, of course, was in periods of rapid economic growth, when the indigenous population found promotion to the higher echelons easy, and their places were taken by newcomers.

Here then we have at last something which seems to be unique to the Northern Ireland labour market operation. There was indeed expansion in the years of growth, the 1960s and early 1970s. Growth meant a double advantage for the country: more workers could find jobs elsewhere, especially in the rest of the UK, in the construction industries and other sectors. At the same time, the operation of a vigorous regional policy led to the establishment of a large number of new industries, especially in the core region. Spending power increased as real incomes rose, and retailing and service industries also flourished.

Why the Boom did not Help

Yet during this classic boom period, the recruitment pattern did not change. Catholics did not obtain a foothold in the traditional bastions of Protestant privilege. When new plants were established, often by foreign-owned firms, and top management was brought in from outside, the situation did not change in favour of the Catholics. Just how this was accomplished, we cannot say. Some of it, no doubt, was 'gate-keeping' on the part of locally recruited Protestant management. Some was due to the fact that the majority of plants were established in areas which were predominantly Protestant. Some of the failure of the Catholics to obtain a foothold may have been due to their reluctance to apply for vacancies in plants where they knew they would be in a minority, in localities where there was only a small Catholic community.

Some Catholics, of course, did find work in the new plants. The analysis of occupational status shows that in 1981, twenty years after the upsurge began,they still did not have high-status jobs in industries even if they were represented in the workforce, albeit not in proportion to their numbers in the total population.

An Explanation?

In these last sections we have tried to adduce every possible reason why Catholics are seriously underrepresented in the workforce, and especially in the better-paid positions. Yet all these do not add up to an explanation. There has to be (and this is admitted even by those who play it down) a residual element of discrimination, or prejudice. As far as the public services are concerned, and in some investigations into identifiable sectors of private firms and corporations, the extent of discrimination has been amply documented. Apart from the *causes célèbres*, like Harland and Wolff and Short Brothers, Protestant hegemony obtained during the years of growth, and it can only have intensified in the years of decline. If Catholics were recruited when there were no Protestants available, they would lose their places under the 'last in, first out' rule which is general practice in industry all over the world.

If we must accept, then, that Catholic disadvantage in the labour market is of long standing, was exacerbated in the period of growth, and accelerated again when decline set in, there are all sorts of consequences flowing from this situation. Catholics remain poor; their spending power is low. A high proportion of them live on state benefits, most of them means-tested. This means that their expenditure within their own communities is relatively low. We have tried to show that, overall, incomes of Northern Irish households are not significantly below those of other UK regions, when taking into account the low cost of housing in particular. Car ownership is fairly high, and so is the ownership of consumer durables generally. This, however, applies in the main to the working population. Those 35 percent of households which contain no earners, whether they consist only of elderly, sick or unemployed, have little purchasing power. Thus service incomes within Catholic communities will also be low: there will be a negative multiplier, as it were.

This general picture is familiar from the studies of the British inner cities, and of the disadvantaged regions, undertaken in the 1970s and 1980s. The 'cycle of disadvantage' must operate not only for families, but for whole communities. This helps to explain the lack of managerial and higher-status jobs in the personal service sector in the Northern Irish area groups which have the characteristics of the peripheral economy. That Catholics are also proportionately underrepresented in the Belfast private service sector cannot be due to lack of purchasing power, except within the Catholic enclave of West Belfast. In the capital city, we have to rely on explanations other than lack of purchasing power.

Developments since 1981

The position we analysed was mainly that obtaining in 1981, for reasons of comparability. Since then unemployment has risen further, by about 30 percent early in 1986 compared with 1981. During this period, more and more qualified Catholic youngsters have come into the labour market, and anti-discrimination legislation has probably been enforced more vigorously than before. On the other hand, jobs in the public sector (where Catholics had in theory the best chance to succeed) are declining, and, with the reduction in births, three of the mainstays of Catholic employment (teaching, nursing, social work) are threatened purely on a proportional basis. Our guess would be that, year by year, the excess unemployment of Catholics will have grown. Unfortunately the Continuous Household

Survey of 1983 cannot be used to make exact comparisons with the 1981 census, but some of the tables suggest that Catholic unemployment was by 1983 two and a half times greater than Protestant unemployment. For example, PPRU (1985), Table 4.2 gives the following combined percentages for 1983–84. All Catholics unemployed: male 35, female 17, all Catholics 28. All Protestants unemployed: male 15, female 11, all Protestants 13. The percentage of Catholics registered at job centres was 13 compared with 5 for Protestants. There were more long-term unemployed, there was more pessimism about chances, and so on: but most of these survey findings have no Census equivalent.

Overview

Catholics in Northern Ireland have an adverse demographic structure under the conditions of a shrinking labour market, exacerbating a disadvantage they have suffered over a long period. We have shown that high fertility in the past is one constituent part of that disadvantage; another is the relatively small number of men, especially, retiring from the labour force. In a 'closed' labour market situation, where entrants mostly succeed leavers of their own denomination, and in the same locality, the low economic activity rates of the older Catholics mean that few vacancies occur through natural wastage. Besides, the older age groups are 'eroded generations' (in the demographic sense) because such a high proportion of men, in particular, emigrated in the post-war years and until the late 1960s.

The geographical distribution of the Catholics is obviously a 'cause' of unemployment, in the sense that the areas where they form a high proportion, or even a majority of the population, are precisely those which, because they are peripheral to the main Northern Irish employment base, have suffered worst in the recession. We have shown that moving out of those areas into the regions of greater opportunity is very difficult; this is probably so objectively, and it is probably even more the case because of the perceived obstacles to a move in search of jobs, and the prevailing pessimism about obtaining work even if vacancies are known to exist, and because of fears associated with moving into a strange environment.

We have shown that recent educational attainments do not demonstrate that Catholics lack the skills required in the labour markets of the 1980s. We have, however, also shown that in the past some bias in the Catholic educational system existed towards the traditional humanities subjects, and whilst this fitted them for employment as teachers, nurses and clerks, it may not have been enough to give them any advantage in the growing advanced service sectors of the 1970s and 1980s.

All this amounts to at least a partial explanation of high Catholic unemployment, low-status occupations, and low household incomes. It is not, however, a sufficient explanation of the persistence of this disadvantage over two decades, when prosperity increased for at any rate twelve or fifteen years; when the NIHE took over housing; when educational opportunities increased, and relevant subjects grew within the Catholic schools' curriculum; and when the laws against discrimination were given a sharper edge.

So we are left with the conclusion that discrimination is not only a residual explanation, but an important component part of the total labour market situation. If people will not move, or apply for jobs, because they fear discrimination in obtaining work and promotion, and because they are anxious about leaving the protection of their community, then perceived hostile and unfair practices on the part of the majority population become a central part of the total explanation of differentials.

The Future

As far as the future of public policies is concerned, it is useless to suggest to Catholics that they should reduce their fertility to improve their chances of finding employment. This is the typical Malthusian reaction to the existence of poverty among workers, and it is as inappropriate now as it was two hundred years ago. Yet the idea is still prevalent in Northern Ireland. Fertility, as we have shown, is falling both among Catholics and among Protestants; for both denominations it is still a good deal higher than the British average, though lower than that of the Republic of Ireland.

Not does it seem appropriate to suggest that more of them should leave the areas where they now live and go to Belfast, where unemployment is rising as swiftly as anywhere else (or more so if we take the latest period). It is equally pointless to suggest that more of them should go to universities, polytechnics, teachers' colleges or colleges of further education. There would not be jobs for them when they graduate or qualify, either in Northern Ireland or elsewhere. Although government training schemes (Northern Ireland Economic Council Report no. 27) seem to be taking some youngsters out of unemployment in Northern Ireland as they do in the UK, they are unlikely to equip many of them with skills which will improve their chances in the long run if traditional recruiting and promotion practices persist. For the bulk of the long-term adult unemployed, the available retraining schemes are inadequate and many of them also irrelevant; but we can safely leave these criticisms to those who have made a more detailed study of them both in Northern Ireland and in Great Britain.

All this leaves, in our view, only two possible lines of advance. First of all, efforts to create additional employment have to be retained and increased. We know that this is a rather different demand in the case of Northern Ireland than in the rest of Great Britain, where the rundown of public expenditure has been much more severe. But even in Northern Ireland there is a great deal more to be done, especially to the housing fabric and the environment generally, as nobody who knows Belfast needs to be told. The rundown of public services has to be halted. Security, at least in the police and private security sectors, is a slightly different matter: nobody would hope that the situation deteriorates again to the point where more police, firemen and security guards must be recruited.

Secondly, and more importantly, the whole area of discrimination against Catholics has to be tackled with greater vigour. It will take another decade at least of improved recruitment policies and promotion procedures to reduce present inequalities significantly – not least because of the unfavourable ratio of labour market entrants compared with the existing workforce.

Fortunately, there is some relief in prospect. The demographic picture will change within the next decade at least at the bottom end: fewer babies have been born since the mid 1960s, and the number is still declining. The number of men approaching retiring age still in employment may be rising slightly, and thus the rate of job vacancies arising from retirement may increase: but this will not be the case if still more men over 50 are forced or persuaded to become economically inactive, which, as we know, normally leads to the elimination of their workplaces rather than to an immediate vacancy for a youngster. However, the present unfavourable ratios between entrants and leavers should automatically improve by the early 1990s – in the rest of the UK as in Northern Ireland.

This natural change, however, will take time. It is not a reason for letting things take their natural course. It is cited here because we feel it should encourage policy-makers and administrators to know that they will work with the tide, not against it.

Public Policy

Apart from this prospect of medium-term population structure improvement, the main constituents of the public policy platform must be the same as those which have been in operation for the last two decades or more:

1 To pursue vigorous regional policies, giving financial encouragement to firms to locate manufacturing activities in Northern Ireland.
2 To keep up a differentially higher rate of public expenditure in Northern Ireland, compared with the poorer regions of Britain, because unemployment is higher and more long-term, and the backlog of environmental damage is greater than anywhere else, and because the segregated nature of settlements means that large communities are without much locally generated earned income.
3 To continue to investigate allegations of illegal practices in industry and the public services: discrimination against the minority in recruiting workers; and failure to promote Catholics to posts of status, responsibility and better pay.

It is not for the outside observer to stress these points. The Northern Ireland Economic Council on the general theme of support for employment, and the Fair Employment Agency in the field of discrimination, have produced a great deal of evidence showing the need to do more of what has already been done in the past; and to continue high levels of expenditure on research and enforcement as well as support, even at a time when public expenditure is being cut back in the rest of the UK.

In recent years, new economic and social research centres have been set up in Northern Ireland. Their detailed investigations are valuable, but have so far thrown no new light on the causes of differential disadvantage. The Continuous Household Survey, after some initial difficulties, has made it possible to observe, in between the decennial Census counts, the changing economic and social structure of the country, often distinguishing between Catholics and 'other denominations'. Some of the Survey findings also make a breakdown possible between urban and rural areas, or even between major regions. Again, though much of this is interesting (especially as regards attitudes, for instance, to moving house), the Continuous Household Survey has so far provided few new insights into the most persistent problem of the Northern Irish economy: that within an overall poor performance, the Catholics have been worse off than the Protestants. Nothing in the Survey suggests that the fears of the minority are not real, and it is a matter of concern that despite the rectification of some serious obstacles to Catholics finding employment, the relative position of the main denominations remains the same.

The efforts made to remove this anomaly are relatively large, when seen against the background of parallel movements in Great Britain to deal with gender and ethnic disadvantage. One could not hope that such efforts would succeed in eradicating, in ten or twenty years, the disadvantages created by centuries of prejudice. That these efforts have so far had only limited success is not a reason for discontinuing them. Now that some of the most obvious barriers to moves towards equality of treatment have been removed (NIHE housing, educational and training activities, enforcement of anti-discrimination legislation), and at a time when the demographic picture seems to be turning in favour of better employment prospects for young people, we should be able to look forward to a decade of success in closing the gap. It remains to be seen whether, now that some of the preconditions have been met, this will actually happen. The 1991 Census should show us whether the situation has improved. If it has not, then the suspicion that discrimination, and the associated fears of the minority, have much to do with the persistence of the differential, must be reinforced.

As far as the situation in the early 1980s is concerned, we have identified this discrimination in the labour market as an important element in the total situation, without, however, being able to quantify it. The evidence required to do this is to be found in the reports of the Fair Employment Agency, and even these cannot tell us how many fewer Catholics would be unemployed if there were no discriminatory practices.

Appendix A

List of Area Groups

Area group I: Belfast
Belfast
Castlereagh
Lisburn
Carrickfergus
Newtownabbey
North Down
Ards

Area group II: Northern
Ballymena
Balleymoney
Coleraine
Larne
Magherafelt
Moyle
Antrim

Area group III: Southern
Cookstown
Craigavon
Armagh
Banbridge
Dungannon
Newry and Mourne
Down

Area group IV: Western
Londonderry
Limavady
Omagh
Strabane
Fermanagh

Appendix B

List of Industry Divisions

0 Agriculture, forestry and fishing

01 Agriculture and horticulture
02 Forestry
03 Fishing

1 Energy and water supply industries

11 Coal extraction and manufacture of solid fuels
12 Coke ovens
13 Extraction of mineral oil and natural gas
14 Mineral oil processing
15 Nuclear fuel production
16 Production and distribution of electricity, gas and other forms of energy
17 Water supply industry

2 Extraction of minerals and ores other than fuels; manufacture of metals, mineral products and chemicals

21 Extraction and preparation of metalliferous ores
22 Metal manufacturing
23 Extraction of minerals not elsewhere specified
24 Manufacture of non-metallic products
25 Chemical industry
26 Production of man-made fibres

3 Metal goods, engineering and vehicle industries

31 Manufacture of metal goods not elsewhere specified
32 Mechanical engineering
33 Manufacture of office machinery and data processing equipment
34 Electrical and electronic engineering
35 Manufacture of motor vehicles and parts thereof
36 Manufacture of other transport equipment
37 Instrument engineering

4 Other manufacturing industries

41, 42 Food, drink and tobacco manufacturing industries
43 Textile industry
44 Manufacture of leather and leather goods
45 Footwear and clothing industries
46 Timber and wooden furniture industries

47 Manufacture of paper and paper products; printing and publishing
48 Processing or rubber and plastics
49 Other manufacturing industries

5 Construction
50 Construction

6 Distribution, hotels and catering; repairs
61 Wholesale distribution (except dealing in scrap and waste materials)
62 Dealing in scrap and waste materials
63 Commission agents
64 }
65 } Retail distribution
66 Hotels and catering
67 Repair of consumer goods and vehicles

7 Transport and communication
71 Railways
72 Other inland transport
74 Sea transport
75 Air transport
76 Supporting services to transport
77 Miscellaneous transport services and storage not elsewhere specified
79 Postal services and telecommunications

8 Banking, finance, insurance, business services and leasing
81 Banking and finance
82 Insurance, except for compulsory social security
83 Business services
84 Renting of movables
85 Owning and dealing in real estate

9 Other services
91 Public administration, national defence and compulsory social security
92 Sanitary services
93 Education
94 Research and development
95 Medical and other health services; veterinary services
96 Other services provided to the general public
97 Recreational services and other cultural services
98 Personal services
99 Domestic services

00 Diplomatic representation, international organizations, allied armed forces
– Industry inadequately described
– Place of work outside United Kingdom

Appendix C

List of Occupational Orders

1 Professional and related supporting managements; senior national and local government managers

001 Judges, barristers, advocates, solicitors
002 Accountants, valuers, finance specialists
003 Personnel and industrial relations managers; O and M, work study and operational research officers
004 Economists, statisticians, systems analysts, computer programmers
005 Marketing, sales, advertising, public relations and purchasing managers
006 Statutory and other inspectors
007 General administrators – national government
008 Local government officers (administrative and executive functions)
009 All other professional and related supporting management and administration

2 Professional and related in education, welfare and health

010 Teachers in higher education
011 Teachers not elsewhere classified
012 Vocational and industrial trainers, education officers, social and behavioural scientists
013 Welfare workers
014 Clergy, ministers of religion
015 Medical and dental practitioners
016 Nurse administrators, nurses
017 Pharmacists, radiographers, therapists not elsewhere classified
018 All other professional and related in education, welfare and health

3 Literary, artistic and sports

019 Authors, writers, journalists
020 Artists, designers, window dressers
021 Actors, musicians, entertainers, stage managers
022 Photographers, cameramen, sound and vision equipment operators
023 All other literary, artistic and sports

4 Professional and related in science, engineering, technology and similar fields

024 Scientists, physicists, mathematicians
025 Civil, structural, municipal, mining and quarrying engineers
026 Mechanical and aeronautical engineers
027 Electrical and electronic engineers
028 Engineers and technologists not elsewhere classified

029 Draughtsmen
030 Laboratory and engineering technicians, technician engineers
031 Architects, town planners, quantity, building and land surveyors
032 Officers (ships and aircraft), air traffic planners and controllers
033 Professional and related in science, engineering and other technologies and similar fields not elsewhere classified

5 Managerial
034 Production, works and maintenance managers, works foremen
035 Site and other managers, agents and clerks of works, general foremen (building and civil engineering)
036 Managers in transport, warehousing, public utilities and mining
037 Office managers
038 Managers in wholesale and retail distribution
039 Managers of hotels, clubs etc. and in entertainment and sport
040 Farmers, horticulturalists, farm managers
041 Officers, UK armed forces
042 Officers, foreign and Commonwealth armed forces
043 Senior police, prison and fire service officers
044 All other managers

6 Clerical and related
045 Supervisors of clerks, civil service executive officers
046 Clerks
047 Retail shop cashiers, checkout and cash-and-wrap operators
048 Supervisors of typists, office machine operators, telephonists etc.
049 Secretaries, shorthand typists, receptionists
050 Office machine operators
051 Telephonists, radio and telegraph operators
052 Supervisors of postmen, mail sorters, messengers
053 Postmen, mail sorters, messengers

7 Selling
054 Sales supervisors
055 Salesmen, sales assistants, shop assistants, shelf fillers, petrol pump, forecourt attendants
056 Roundsmen, van salesmen
057 Sales representatives and agents

8 Security and protective service
058 NCOs and other ranks, UK armed forces
059 NCOs and other ranks, foreign and Commonwealth armed forces
060 Supervisors (police sergeants, fire fighting and related)
061 Policemen, firemen, prison officers
062 Other security and protective service workers

9 Catering, cleaning, hairdressing and other personal services
063 Catering supervisors
064 Chefs, cooks
065 Waiters and bar staff

066 Counter hands, assistants, kitchen porters, hands
067 Supervisors – housekeeping and related
068 Domestic staff and school helpers
069 Travel stewards and attendants, hospital and hotel porters
070 Ambulancemen, hospital orderlies
071 Supervisors, foremen – caretaking, cleaning and related
072 Caretakers, road sweepers and other cleaners
073 Hairdressing supervisors
074 Hairdressers, barbers
075 All other in catering, cleaning and other personal service

10 Farming, fishing and related
076 Foremen – farming, horticulture, forestry
077 Farm workers
078 Horticulture workers, gardeners, groundsmen
079 Agricultural machinery drivers, operators
080 Forestry workers
081 Supervisors, mates – fishing
082 Fishermen
083 All other in farming and related

11 Materials processing; making and repairing (excluding metal and electrical)
084 Foremen – tannery and leather (including leather substitutes) working
085 Tannery and leather (including leather substitutes) workers
086 Foremen – textile processing
087 Textile workers
088 Foremen – chemical processing
089 Chemical, gas and petroleum process plant operators
090 Foremen – food and drink processing
091 Bakers, flour confectioners
092 Butchers
093 Foremen – paper and board making and paper products
094 Paper, board and paper product makers, bookbinders
095 Foremen – glass, ceramics, rubber, plastics etc.
096 Glass, ceramics, furnacemen and workers
097 Rubber and plastics workers
098 All other in processing materials (other than metal)
099 Foremen – printing
100 Printing workers, screen and block printers
101 Foremen – textile materials working
102 Tailors, dressmakers and other clothing workers
103 Coach trimmers, upholsterers, mattress makers
104 Foremen – woodworking
105 Woodworkers, pattern makers
106 Sawyers, veneer cutters, woodworking machinists
107 All other in making and repairing (excluding metal and electrical)

12 Processing, making, repairing and related (metal and electrical)
108 Foremen – metal making and treating

109 Furnacemen (metal), rollermen, smiths, forgemen
110 Metal drawers, moulders, die casters, electroplaters, annealers
111 Foremen – engineering machining
112 Press and machine tool setter operators and operators, turners
113 Machine attendants, minders, press and stamping machine operators, metal polishers, fettlers, dressers
114 Foremen – production fitting (metal)
115 Tool makers, tool fitters, markers-out
116 Instrument and watch and clock makers and repairers
117 Metal working production fitters and fitter/machinists
118 Motor vehicle and aircraft mechanics
119 Office machinery mechanics
120 Foremen – production fitting and wiring (electrical)
121 Production fitters, electricians, electricity power plant operators, switchboard attendants
122 Telephone fitters, cable jointers, linesmen
123 Radio, TV and other electronic maintenance fitters and mechanics
124 Foremen – metal working, pipes, sheets, structures
125 Plumbers, heating and ventilating fitters, gas fitters
126 Sheet metal workers, platers, shipwrights, riveters etc.
127 Steel erectors, scaffolders, steel benders, fixers
128 Welders
129 Foremen – other processing, making and repairing (metal and electrical)
130 Goldsmiths, silversmiths etc., engravers, etchers
131 All other in processing, making and repairing (metal and electrical)

13 Painting, repetitive assembling, product inspecting, packaging and related
132 Foremen – painting and similar coating
133 Painters, decorators, french polishers
134 Foremen – product assembling (repetitive)
135 Repetitive assemblers (metal and electrical goods)
136 Foremen – product inspection and packaging
137 Inspectors, viewers, testers, packers, bottlers etc.
138 All other in painting, repetitive assembling, product inspection, packaging and related

14 Construction, mining and related not identified elsewhere
139 Foremen – building and civil engineering not elsewhere classified
140 Building and construction workers
141 Concreters, road surfacers, railway lengthmen
142 Sewage plant attendants, sewermen (maintenance), mains and service layers, pipe jointers (gas, water, drainage, oil), inspectors (water supply), turncocks
143 Civil engineering labourers, craftmen's mates and other builders' labourers not elsewhere classified
144 Foremen/deputies – coal mining
145 Face-trained coal mining workers
146 All other in construction, mining, quarrying, well drilling and related not elsewhere classified

15 Transport operating, materials moving and storing and related
147 Foremen – ships, lighters and other vessels
148 Deck, engine-room hands, bargemen, lightermen, boatmen
149 Foremen – rail transport operating
150 Rail transport operating staff
151 Foremen – road transport operating, bus inspectors
152 Bus, coach, lorry drivers etc.
153 Bus conductors, drivers' mates
154 Foremen – civil engineering plant operating, materials handling equipment
155 Mechanical plant, fork lift, mechanical truck drivers, crane drivers, operators
156 Foremen – materials moving and storing
157 Storekeepers, stevedores, warehouses, market and other goods porters
158 All other in transport operating, materials moving and storing and related not elsewhere classified

16 Miscellaneous
159 Foremen – miscellaneous
160 General labourers
161 All other in miscellaneous occupations not elsewhere classified

17 Inadequately described and not stated
– Inadequately described occupations
– Occupations not stated

Appendix D

List of Additional Tables

Population

Migration

Development of the labour market

Unemployment

Employment status and religion

Industrial structure

Bibliography

Unpublished statistical sources are not included in this bibliography.

Birrell, W.D., Hillyard, P.A.R., Murie, A. and Roche, D.J.D. (1971) *Housing in Northern Ireland*. University Working Paper 12. London: CES.

Boal, F.N. (1981) 'Residential Segregation and Mixing in a Situation of Ethnic and National Conflict: Belfast', in P.A. Compton (ed.), *The Contemporary Population of Northern Ireland and Population Related Issues*. Belfast: The Queen's University.

Boal, F., Murray, R.C. and Poole, M.A. (1976) 'Belfast, the Urban Encapsulation of a National Conflict', in S.E. Clarke and J.E. Obler (eds), *Urban Ethnic Conflict: a Comparative Perspective*. Chapel Hill, NC: Department for Research in Social Science, University of North Carolina.

Bradley, J., Hewitt, V.N. and Jefferson, C.W. (1984) *Industrial Location Policy and Equality of Opportunity in Assisted Employment in Northern Ireland, 1949–81*. Unpublished. Belfast: The Queen's University.

Central Statistical Office (1983) *Annual Abstract of Statistics 1982*. London: HMSO.

Central Statistical Office (1986) *Annual Abstract of Statistics 1985*. London: HMSO.

Compton, P.A. (1974) 'Northern Irishmen Stay at Home', *Geographical Magazine*, 44: 254–6.

Compton, P.A. (1981) (ed.) *The Contemporary Population of Northern Ireland and Population Related Issues*. Belfast: The Queen's University.

Compton, P.A. (1985a) 'The 1981 Northern Ireland Census of Population: Estimates of Non-Enumerated Population', in C. Morris, P. Compton and A. Luke (eds), *Non-Enumeration in the 1981 Northern Ireland Census of Population*. PPRU Occasional Paper no. 9. Belfast: PPRU.

Compton, P.A. (1985b) 'An Evaluation of the Changing Religious Composition of the Population of Northern Ireland', *Economic and Social Review*, 16 (3): 201–24.

Compton, P.A., Coward, J. and Wilson-Davis, K. (1985) 'Family Size and Religious Denomination in Northern Ireland', *Journal of Biosocial Science*, 17: 137–45.

Compton, P.A. and Power, J.P. (1986) 'Estimates of the Religious Composition of Northern Ireland Local Government Districts in 1981 and Change in the Geographical Pattern of Religious Composition between 1971 and 1981', *Economic and Social Review*, 17 (2): 87–105.

Coopers and Lybrand (1982) *The Northern Ireland Economy, the Current Economic Situation and Prospects, Mid-Year Review*. Northern Ireland: Coopers and Lybrand.

Cormack, R.J. and Osborne, R.D. (eds) (1983) *Religion, Education and Employment: Aspects of Equal Opportunity in Northern Ireland*. Belfast: Appletree Press.

Cormack, R.J., Osborne, R.D. Reid, N.G. and Williamson, A.P. (1984) *Participation in Higher Education: Trends in the Social and Spatial Mobility of Northern Ireland Undergraduates*. Belfast: SSRC.

Cormack, R.J., Osborne, R.D. and Thompson, W.T. (1980) *Into Work? Young School Leavers and the Structure of Opportunity in Belfast*. Research Paper 5. Belfast: Fair Employment Agency.

Cullen, L.M. (1972) *An Economic History of Ireland since 1660*. London: Batsford.

Department of Economic Development, Northern Ireland (1984) *Labour Force Survey 1981*. Belfast: DED.

Department of Education for Northern Ireland (1982) *Directory of Vocational Courses for 1982/83 (including Agricultural Courses)*. Belfast: DENI.

Department of Education for Northern Ireland (1984) *Statistical Bulletin* no. 2. Belfast: DENI.

Department of Employment (1985) *Family Expenditure Survey 1983*. Belfast: HMSO.

Doherty, P. (1980) 'Patterns of Unemployment in Belfast', *Irish Geography*, 13: 65–76.

Ermisch, J. (1982) 'The Labour Market: Historical Development and Hypotheses', in David Eversley and Wolfgang Köllman (eds), *Population Change and Social Planning*. London: Edward Arnold.

Eversley, D. and Herr, V. (1985) *Roman Catholic Population of Northern Ireland in 1981: a Revised Estimate*. Belfast: Fair Employment Agency.

Fair Employment Agency (1978) *An Industrial and Occupational Profile of the Two Sections of the Community in Northern Ireland*. Belfast: FEA.

Fair Employment Agency (1983) *Report of an Investigation by the Fair Employment Agency for Northern Ireland into the Non-Industrial Northern Ireland Civil Service*. Belfast: FEA.

Garvey, D. (1985) 'The History of Migration Flows in the Republic of Ireland: A Review of Patterns and Trends in Migration, especially between Ireland and the United Kingdom', *Population Trends*, 39.

General Register Office (1984) *Northern Ireland Annual Report 1981*. Belfast: HMSO.

Great Britain Census 1981, *National Migration Report*, Part I. London: HMSO.

Halifax Building Society (1986) *Regional Bulletin*, no. 10.

Harrison, R.L. (1981) 'Population Change and Housing in Belfast', in P.A. Compton (ed.) *The Contemporary Population of Northern Ireland and Population Related Issues*. Belfast: The Queen's University.

Harrison. R.T. (1980) *Regional Planning and Employment Location Trends in Northern Ireland*. Unpublished. Belfast: The Queen's University.

Hepburn, A.C. (1982) *Employment and Religion in Belfast 1901–1970*. Belfast: FEA.

Jackson. J.A. (1963) *The Irish in Britain*. London: Routledge and Kegan Paul.

Lane, M. and Thompson, J. (1985) 'Immigrants Ten Years on: Continuities and Change', in *Measuring Sociodemographic Change*. British Society for Population Studies Occasional Paper 34. London: OPCS.

McCartney, J.R. and Whyte, J. (1984) *Opportunity and Choice. Part One: Fifteen Year Olds in Schools and Further Education Colleges. Part Two: Sixteen–Nineteen Year Olds in Full Time Courses in Schools and Further Education Colleges*. Reports 28 and 29. Belfast: Northern Ireland Council for Educational Research.

Miller, R. (1979) *Occupational Mobility of Protestants and Roman Catholics in Northern Ireland*. Research Paper 4. Belfast: FEA.

Miller, R.L. and Osborne, R.D. (1983) 'Religion and Unemployment: Evidence from a Cohort Survey', in R.J. Cormack and R.D. Osborne (eds), *Religion, Education and Employment: Aspects of Equal Opportunity in Northern Ireland*. Belfast: Appletree Press.

Morris, C. and Wilson-Davis, K. (1983) *Family Expenditure Survey Report for 1981*. Belfast: PPRU.

Morris, C., Compton, P. and Lake, A. (1985) *Non-Enumeration in the 1981 Northern Ireland Census of Population*. PPRU Occasional Paper no. 9. Belfast, PPRU.

Murray, D. and Darby, J. (1980) *The Vocational Aspirations and Expectations of School Leavers in Londonderry and Strabane*. Belfast: FEA.

Northern Ireland *Annual Abstract of Statistics* no. 1, 1981. Belfast: HMSO, 1982.

Northern Ireland *Annual Abstract of Statistics* no. 2, 1982. Belfast: HMSO, 1983.

Northern Ireland *Annual Abstract of Statistics* no. 3, 1983. Belfast: HMSO, 1984.

Northern Ireland Census 1971, *Religion Tables* (1975). Belfast: HMSO.

Northern Ireland Census 1971, *Summary Report* (1973). Belfast: HMSO.

Northern Ireland Census 1981, *Preliminary Report* (1982). Belfast: HMSO.

Northern Ireland Census 1981, *Economic Activity Report* (1983). Belfast: HMSO.

Northern Ireland Census 1981, *Migration Report* (1983). Belfast: HMSO.

Northern Ireland Census 1981, *Religion Report* (1984). Belfast: HMSO.

Northern Ireland Census 1981, *Summary Report* (1983). Belfast: HMSO.

Northern Ireland Census 1981, *Workplace and Transport to Work Report* (1983). Belfast: HMSO.

Northern Ireland Economic Council Reports:
No. 23, *Employment Patterns in Northern Ireland 1950–80* (June 1981)
No. 27, *Youth Employment and Training* (March 1982)
No. 38, *Economic Strategy: Historical Growth Performance* (May 1983)
No. 45, *Textile Industry* (November 1984)
No. 46, *Engineering Industry* (December 1984)
No. 48, *Review of Recent Development in Housing Policy* (February 1985)
No. 51, *Food, Drink and Tobacco Industries* (May 1985)
No. 57, *Demographic Trends in Northern Ireland* (April 1986)
Belfast: Northern Ireland Economic Council.

Northern Ireland Economic Council (1985) *Higher Education in Ireland: Co-operation and Complementarity*. Dublin and Belfast: the National Economic and Social Council and the Northern Ireland Economic Development Office.

Northern Ireland Housing Executive (May–Oct. 1978) *Household Movement from Belfast*. Unpublished.

Northern Ireland Housing Executive (1983) *The Public Sector and the Private Housing Market*. Unpublished.

Northern Ireland Housing Executive (no date) *Demographic Change and Future Housing Need in Northern Ireland 1981–1987*. Unpublished.

Northern Ireland Housing Executive (1981–83) Regional Reports and District Analyses, *Regional Household Survey 1981*. Unpublished.

Northern Ireland Housing Executive (1982) *Northern Ireland Public Sector Housebuilding Programme 1981–82 to 1985–86*. Unpublished.

Office of Population Censuses and Surveys (OPCS) (1975ff.) *International Migration Annual Reports*. Series MN. London: HMSO.

Office of Population Censuses and Surveys and Registrar General Scotland (1983) *Census 1981, Country of Birth; Great Britain*. London: HMSO.

Office of Population Censuses and Surveys *Monitors, Recorded Internal Population Movement in the United Kingdom*. MN83/4(1983), MN84/4(1984), MN85/4(1985a). London: HMSO.

Office of Population Censuses and Surveys *International Migration: 1983*. Series MN no.10. London HMSO: 1985.

Osborne, R.D. (1984) *Qualifications of School Leavers by Travel To Work Areas (1982), Girls, by Religion*. Unpublished. Jordanstown: Ulster Polytechnic.

Osborne, R.D. (1985) *Religion and Educational Qualifications in Northern Ireland*. Research Paper 8. Belfast: Fair Employment Agency.

Osborne, R.D. (no date) *Segregated Schools and Examination Results in Northern Ireland: Some Preliminary Research*. Unpublished. Jordanstown: Ulster Polytechnic.

Osborne, R.D. and Cormack, R.J. (1986) 'Unemployment and Religion in Northern Ireland', *Economic and Social Review*, 17: 215–25.

Osborne, R.D. and Cormack, R.J. (1987) *Religion, Occupations and Employment 1971–81*. Research Paper 11. Belfast: FEA.

Osborne, R.D., Cormack, R.J., Reid, N.G. and Williamson, A.P. (1983) 'Political Arithmetic, Higher Education and Religion in Northern Ireland', in R.J. Cormack and R.D. Osborne (eds), *Religion, Education and Employment: Aspects of Equal Opportunity in Northern Ireland*. Belfast: Appletree Press.

Osborne, R.D. and Murray, R.C. (1978) *Educational Qualifications and Religious Affiliation in Northern Ireland*. Research Paper 3. Belfast: FEA.

Patton, J. (1981) *Directional Bias in Inter-Urban Migration in the Public Housing Sector in Belfast*. Unpublished PhD thesis. Jordanstown: Ulster Polytechnic.

Policy Planning Research Unit (PPRU) *Monitors* Continuous Household Survey. No. 1/84 (1984), no. 2/85 (1985), no. 1/86 (1986). Belfast: PPRU.

Policy Planning Research Unit (1986) *Revision of Estimate of Non-Enumeration at 1981 Northern Ireland Census of Population* (with associated adjustments). SCG(86)2. Belfast: PPRU.

Regional Trends no. 22 (1987). London: HMSO.

Singleton, D. (1981) 'Planning Implications of Population Trends in Northern Ireland', in P.A. Compton (ed.), *The Contemporary Population of Northern Ireland and Population Related Issues*. Belfast: The Queen's University.

Index